Birkhäuser

Trends in Mathematics

Trends in Mathematics is a series devoted to the publication of volumes arising from conferences and lecture series focusing on a particular topic from any area of mathematics. Its aim is to make current developments available to the community as rapidly as possible without compromise to quality and to archive these for reference.

Proposals for volumes can be submitted using the Online Book Project Submission Form at our website www.birkhauser-science.com.

Material submitted for publication must be screened and prepared as follows:

All contributions should undergo a reviewing process similar to that carried out by journals and be checked for correct use of language which, as a rule, is English. Articles without proofs, or which do not contain any significantly new results, should be rejected. High quality survey papers, however, are welcome.

We expect the organizers to deliver manuscripts in a form that is essentially ready for direct reproduction. Any version of TEX is acceptable , but the entire collection of files must be in one particular dialect of TEX and unified according to simple instructions available from Birkhäuser.

Furthermore, in order to guarantee the timely appearance of the proceedings it is essential that the final version of the entire material be submitted no later than one year after the conference.

More information about this series at http://www.springer.com/series/4961

Valentin Bertsch • Armin Ardone •
Michael Suriyah • Wolf Fichtner •
Thomas Leibfried • Vincent Heuveline
Editors

Advances in Energy System Optimization

Proceedings of the 2nd International Symposium on Energy System Optimization

OPEN ACCESS

Birkhäuser

Editors
Valentin Bertsch
Department of Energy Systems Analysis
Institute of Engineering Thermodynamics
German Aerospace Center
Stuttgart, Germany

Michael Suriyah
Institute of Electric Energy Systems
and High-Voltage Technology
Karlsruhe Institute of Technology
Karlsruhe, Germany

Thomas Leibfried
Institute of Electric Energy Systems
and High-Voltage Technology
Karlsruhe Institute of Technology
Karlsruhe, Germany

Armin Ardone
Institute for Industrial Production
Karlsruhe Institute of Technology
Karlsruhe, Germany

Wolf Fichtner
Institute for Industrial Production
Karlsruhe Institute of Technology
Karlsruhe, Germany

Vincent Heuveline
Engineering Mathematics and Computing
Lab, Interdisciplinary Center for Scientific
Computing
Heidelberg University
Heidelberg, Germany

ISSN 2297-0215 ISSN 2297-024X (electronic)
Trends in Mathematics
ISBN 978-3-030-32156-7 ISBN 978-3-030-32157-4 (eBook)
https://doi.org/10.1007/978-3-030-32157-4

Mathematics Subject Classification (2010): 90-06, 90-08, 90B90, 90B99, 90B10, 90C05, 90C06, 90C10, 90C11, 90C15, 90C20, 90C29, 90C30, 93A14, 93A15, 93A30

This book is an open access publication.

This book is published under the imprint Birkhäuser, www.birkhauser-science.com, by the registered company Springer Nature Switzerland AG.
The registered company address is: Gewerbestrasse 11, 6330 Cham, Switzerland

Preface

This volume on *Advances in Energy System Optimization* contains a selection of peer-reviewed papers related to the *2nd International Symposium on Energy System Optimization (ISESO 2018).* The symposium was held at Karlsruhe Institute of Technology (KIT) under the symposium theme "Bridging the Gap Between Mathematical Modelling and Policy Support" on October 10–11, 2018. ISESO 2018 was organized by KIT (Institute for Industrial Production (IIP) and Institute of Electric Energy Systems and High-Voltage Technology (IEH)), the Heidelberg Institute for Theoretical Studies (HITS), Heidelberg University (Engineering Mathematics and Computing Lab (EMCL)), German Aerospace Center (DLR, Institute of Engineering Thermodynamics, Department of Energy Systems Analysis), and the University of Stuttgart (Institute for Building Energetics, Thermotechnology, and Energy Storage).

The organizing institutes are engaged in a number of collaborative research activities aimed at combining interdisciplinary perspectives from mathematics, operational research, energy economics, and electrical engineering to develop new approaches to integrated energy system and grid modeling designed to solve real-world energy problems efficiently. The symposium was now held for the second time. By design, ISESO is limited in size as to ensure a productive atmosphere for discussion and reflection on how to tackle the many challenges facing today's and tomorrow's energy systems. Around 50 international participants attended 17 international presentations from both industry and academia including 2 keynote presentations and 15 contributed papers in 6 sessions. The sessions focused on diverse challenges in energy systems, ranging from operational to investment planning problems, from market economics to technical and environmental considerations, from distribution grids to transmission grids, and from theoretical considerations to data provision concerns and applied case studies. The presentations were complemented by a panel discussion on the symposium theme "Bridging the Gap Between Mathematical Modelling and Policy Support" involving senior experts from academia and industry.

The papers in this volume are broadly structured according to the sessions within the symposium as outlined below:

- Optimal Power Flow
- Energy System Integration
- Demand Response
- Planning and Operation of Distribution Grids.

The editors of this volume served as the organizing committee. We wish to thank all the reviewers as well as all the individuals and institutions who worked hard, often invisibly, for their tremendous support. In particular, we wish to thank Nico Meyer-Hübner and Nils Schween for the coordination of the local organization. Finally, we also wish to thank all the participants, speakers, and panelists for their contributions to making ISESO a success.

Stuttgart, Germany — Valentin Bertsch
Karlsruhe, Germany — Armin Ardone
Karlsruhe, Germany — Michael Suriyah
Karlsruhe, Germany — Wolf Fichtner
Karlsruhe, Germany — Thomas Leibfried
Heidelberg, Germany — Vincent Heuveline

Contents

Part I
Optimal Power Flow

Feasibility vs. Optimality in Distributed AC OPF: A Case Study Considering ADMM and ALADIN

Alexander Engelmann and Timm Faulwasser

Abstract This paper investigates the role of feasible initial guesses and large consensus-violation penalization in distributed optimization for Optimal Power Flow (OPF) problems. Specifically, we discuss the behavior of the Alternating Direction of Multipliers Method (ADMM). We show that in case of large consensus-violation penalization ADMM might exhibit slow progress. We support this observation by an analysis of the algorithmic properties of ADMM. Furthermore, we illustrate our findings considering the IEEE 57 bus system and we draw upon a comparison of ADMM and the Augmented Lagrangian Alternating Direction Inexact Newton (ALADIN) method.

Keywords Distributed optimal power flow · ALADIN · ADMM

1 Introduction

Distributed optimization algorithms for AC Optimal Power Flow (OPF) recently gained significant interest as these problems are inherently non-convex and often large-scale; i.e. comprising up to several thousand buses [1]. Distributed

This work is part of a project that receives funding from the European Union's Horizon 2020 research and innovation program under grant agreement No. 730936. TF acknowledges further support from the Baden-Württemberg Stiftung under the Elite Programme for Postdocs.

A. Engelmann (✉)
Institute for Automation and Applied Informatics (IAI), Karlsruhe Institute of Technology (KIT), Eggenstein-Leopoldshafen, Germany
e-mail: alexander.engelmann@kit.edu

T. Faulwasser
Institute for Energy Systems, Energy Efficiency and Energy Economics, TU Dortmund, Dortmund, Germany
e-mail: timm.faulwasser@ieee.org

V. Bertsch et al. (eds.), *Advances in Energy System Optimization*, Trends in Mathematics, https://doi.org/10.1007/978-3-030-32157-4_1

optimization is considered to be helpful as it allows splitting large OPF problems into several smaller subproblems; thus reducing complexity and avoiding the exchange of full grid models between subsystems. We refer to [2] for a recent overview of distributed optimization and control approaches in power systems.

One frequently discussed *convex* distributed optimization method is the Alternating Direction of Multipliers Method (ADMM) [3], which is also applied in context of AC OPF [1, 4, 5]. ADMM often yields promising results even for large-scale power systems [4]. However, ADMM sometimes requires a specific partitioning technique and/or a feasible initialization in combination with high consensus-violation penalization parameters to converge [1]. The requirement of feasible initialization seems quite limiting as it requires solving a centralized inequality-constrained power flow problem requiring full topology and load information leading to approximately the same complexity as full OPF.

In previous works [6–8] we suggested applying the Augmented Lagrangian Alternating Direction Inexact Newton (ALADIN) method to stochastic and deterministic OPF problems ranging from 5 to 300 buses. In case a certain line-search is applied [9], ALADIN provides global convergence guarantees to local minimizers for non-convex problems without the need of feasible initialization. The results in [6] underpin that ALADIN is able to outperform ADMM in many cases. This comes at cost of a higher per-step information exchange compared with ADMM and a more complicated coordination step, cf. [6].

In this paper we investigate the interplay of feasible initialization with high penalization for consensus violation in ADMM for distributed AC OPF. We illustrate our findings on the IEEE 57-bus system. Furthermore, we compare the convergence behavior of ADMM to ALADIN not suffering from the practical need for feasible initialization [9]. Finally, we provide theoretical results supporting our numerical observations for the convergence behavior of ADMM.

The paper is organized as follows: In Sect. 2 we briefly recap ADMM and ALADIN including their convergence properties. Section 3 shows numerical results for the IEEE 57-bus system focusing on the influence of the penalization parameter ρ on the convergence behavior of ADMM. Section 4 presents an analysis of the interplay between high penalization and a feasible initialization.

2 ALADIN and ADMM

For distributed optimization, OPF problems are often formulated in *affinely coupled separable form*

$$\min_{x\in\mathbb{R}^{n_x}} \sum_{i\in\mathcal{R}} f_i(x_i) \quad \text{subject to} \quad x_i \in \mathcal{X}_i,\ \forall i \in \mathcal{R} \quad \text{and} \quad \sum_{i\in\mathcal{R}} A_i x_i = 0, \tag{1}$$

where the decision vector is divided into sub-vectors $x^\top = [x_1^\top, \dots, x_{|\mathcal{R}|}^\top] \in \mathbb{R}^{n_x}$, $\mathcal{R}$ is the index set of subsystems usually representing geographical areas of a power system and local nonlinear constraint sets $\mathcal{X}_i := \{x_i \in \mathbb{R}^{n_{xi}} \mid h_i(x_i) \leq 0\}$.

Algorithm 1 ADMM

Initialization: Initial guesses z_i^0, λ_i^0 for all $i \in \mathcal{R}$, parameter ρ.
Repeat (until convergence):

1. *Parallelizable Step:* Solve for all $i \in \mathcal{R}$ locally

$$x_i^k = \underset{x_i}{\text{argmin}} \quad f_i(x_i) + (\lambda_i^k)^\top A_i x_i + \frac{\rho}{2} \left\| A_i(x_i - z_i^k) \right\|_2^2 \quad \text{s.t.} \quad h_i(x_i) \leq 0. \tag{2}$$

2. *Update dual variables:*

$$\lambda_i^{k+1} = \lambda_i^k + \rho A_i(x_i^k - z_i^k). \tag{3}$$

3. *Consensus Step:* Solve the coordination problem

$$\min_{\Delta x} \sum_{i \in \mathcal{R}} \tfrac{\rho}{2} \Delta x_i^\top A_i^\top A_i \Delta x_i + \lambda_i^{k+1\top} A_i \Delta x_i \tag{4a}$$

$$\text{s.t.} \quad \sum_{i \in \mathcal{R}} A_i(x_i^k + \Delta x_i) = 0. \tag{4b}$$

4. *Update variables:* $z^{k+1} \leftarrow x^k + \Delta x^k$.

Throughout this work we assume that f_i and h_i are twice continuously differentiable and that all $\mathcal{X}_i$ are compact. Note that the objective functions $f_i : \mathbb{R}^{n_{xi}} \to \mathbb{R}$ and nonlinear inequality constraints $h_i : \mathbb{R}^{n_{xi}} \to \mathbb{R}^{n_{hi}}$ only depend on x_i and that coupling between them takes place in the affine consensus constraint $\sum_{i \in \mathcal{R}} A_i x_i = 0$ only. There are several ways of formulating OPF problems in form of (1) differing in the coupling variables and the type of the power flow equations (polar or rectangular), cf. [4, 6, 10].

Here, we are interested in solving Problem (1) via ADMM and ALADIN summarized in Algorithms 1 and 2 respectively.[1,2] At first glance it is apparent that ADMM and ALADIN share several features. For example, in Step (1) of both algorithms, local augmented Lagrangians subject to local nonlinear inequality constraints h_i are minimized in parallel.[3] Observe that while ADMM maintains multiple local Lagrange multipliers λ_i, ALADIN considers one global Lagrange multiplier vector λ. In Step (2), ALADIN computes sensitivities B_i, g_i and C_i (which often can directly be obtained from the local numerical solver without additional computation) whereas ADMM updates the multiplier vectors λ_i.

[1]We remark that there exist a variety of variants of ADMM, cf. [3, 11]. Here, we choose the formulation from [9] in order to highlight similarities between ADMM and ALADIN.

[2]Note that, due to space limitations, we describe the full-step variant of ALADIN here. To obtain convergence guarantees from a remote starting point, a globalization strategy is necessary, cf. [9].

[3]For notational simplicity, we only consider nonlinear inequality constraints here. Nonlinear equality constraints g_i can be incorporated via a reformulation in terms of two inequality constraints, i.e. $0 \leq g_i(x_i) \leq 0$.

Algorithm 2 ALADIN (full-step variant)

Initialization: Initial guess (z^0, λ^0), parameters $\Sigma_i \succ 0, \rho, \mu$.
Repeat (until convergence):

1. *Parallelizable Step:* Solve for all $i \in \mathcal{R}$ locally

$$x_i^k = \underset{x_i}{\operatorname{argmin}} \quad f_i(x_i) + (\lambda^k)^\top A_i x_i + \frac{\rho}{2} \left\| x_i - z_i^k \right\|_{\Sigma_i}^2 \quad \text{s.t.} \quad h_i(x_i) \leq 0 \quad | \, \kappa_i^k.$$

2. *Compute sensitivities:* Compute Hessian approximations B_i^k, gradients g_i^k and Jacobians of the active constraints C_i^k, cf. [9].
3. *Consensus Step:* Solve the coordination problem

$$\begin{aligned} \min_{\Delta x, s} \; & \sum_{i \in \mathcal{R}} \left\{ \tfrac{1}{2} \Delta x_i^\top B_i^k \Delta x_i + {g_i^k}^\top \Delta x_i \right\} + {\lambda^k}^\top s + \tfrac{\mu}{2} \|s\|_2^2 \\ \text{s.t.} \; & \sum_{i \in \mathcal{R}} A_i (x_i^k + \Delta x_i) = s \quad | \, \lambda^{\mathrm{QP}} \\ & C_i^k \Delta x_i = 0 \qquad \forall i \in \mathcal{R}. \end{aligned} \tag{5}$$

4. *Update variables:* $z^{k+1} \leftarrow x^k + \Delta x^k, \quad \lambda^{k+1} \leftarrow \lambda^{\mathrm{QP}}$.
Similarity to ADMM: Remove C_i^k in (5), set $B_i^k = \rho A_i^\top A_i$, $g_i^k = A_i^\top \lambda^k$ and $\Sigma_i = A_i^\top A_i$, set $\mu = \infty$ (i.e. $s = 0$) and use dual ascent step (3) for updating λ^k in (4), cf. [9].

In Step (3), both algorithms communicate certain information to a central entity which then solves a (usually centralized) coordination quadratic program. However, ALADIN and ADMM differ in the amount of exchanged information: Whereas ADMM only communicates the local primal and dual variables x_i and λ_i, ALADIN additionally communicates sensitivities. This is a considerable amount of extra information compared with ADMM. However, there exist methods to reduce the amount of exchanged information and bounds on the information exchange are given in [6]. Another important difference is the computational complexity of the coordination step. In many cases, the coordination step in ADMM can be reduced to an averaging step based on neighborhood communication only [3], whereas in ALADIN the coordination step involves the centralized solution of an equality constrained quadratic program.

In the last step, ADMM updates the primal variables z_i, while ALADIN additionally updates the dual variables λ. Differences of ALADIN and ADMM also show up in the convergence speed and their theoretical convergence guarantees: Whereas ALADIN guarantees global convergence and quadratic local convergence for non-convex problems if a certain globalization strategy is applied [9], few results exist for ADMM in the non-convex setting. Recent works [12, 13] investigate the convergence of ADMM for special classes of non-convex problems; however, to the best of our knowledge OPF problems do not belong to these classes.

3 Numerical Results

Next, we investigate the behavior of ADMM for large ρ and a feasible initialization to illustrate performance differences between ADMM and ALADIN. We consider power flow equations in polar form with coupling in active/reactive power and voltage angle and magnitude at the boundary between two neighbored regions [6]. We consider the IEEE 57-bus system with data from MATPOWER and partitioning as in [6] as numerical test case.

Figures 1 and 2 show the convergence behavior of ADMM (with and without feasible initialization (f.)) and ALADIN for several convergence indicators and two different penalty parameters $\rho = 10^4$ and $\rho = 10^6$.[4] Therein, the left-handed plot depicts the consensus gap $\|Ax^k\|_\infty$ representing the maximum mismatch of coupling variables (active/reactive power and voltage magnitude/angle) at borders between two neighbored regions. The second plot shows the objective function value f^k and the third plot presents the distance to the minimizer $\|x^k - x^\star\|_\infty$ over the iteration index k. The right-handed figure shows the nonlinear constraint violation $\|g(z^k)\|_\infty$ after the consensus steps (4) and (5) of Algorithms 1 and 2 respectively representing the maximum violation of the power flow equations.[5]

In case of small $\rho = 10^4$, both ADMM variants behave similar and converge slowly towards the optimal solution with slow decrease in consensus violation, nonlinear constraint violation and objective function value. If we increase ρ to $\rho = 10^6$ with results shown in Fig. 2, the consensus violation $\|Ax^k\|_\infty$ gets smaller in ADMM with feasible initialization. The reason is that a large ρ forces x_i^k being close to z_i^k leading to small $\|Ax^k\|_\infty$ as we have $Az^k = 0$ from the consensus step. But, at the same time, this also leads to a slower progress in optimality f^k compared to $\rho = 10^4$, cf. the second plot in Figs. 1 and 2.

On the other hand, this statement does not hold for ADMM with infeasible initialization (blue lines in Figs. 1 and 2) as the constraints in the local step

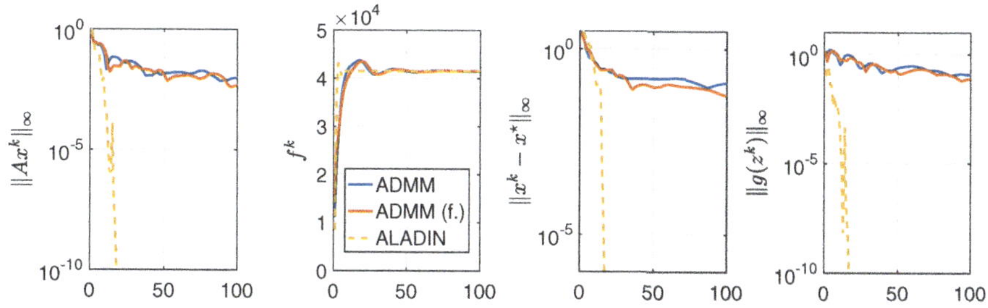

Fig. 1 Convergence behavior of ADMM with infeasible initialization, ADMM with feasible initialization (f.) for $\rho = 10^4$ and ALADIN

[4]The scaling matrices Σ_i are diagonal. They are chosen to improve convergence. Hence, entries corresponding to voltages and phase angles are 100, entries corresponding to powers are set to 1.

[5]The power flow equations for the IEEE 57-bus systems are considered as nonlinear equality constraints $g_i(x_i) = 0$. Hence, $g_i(x_i) \neq 0$ represents a violation of the power flow equations.

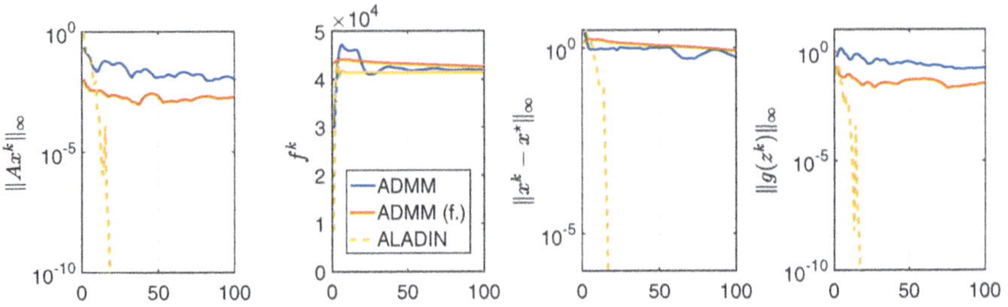

Fig. 2 Convergence behavior of ADMM with infeasible initialization, ADMM with feasible initialization (f.) for $\rho = 10^6$ and ALADIN

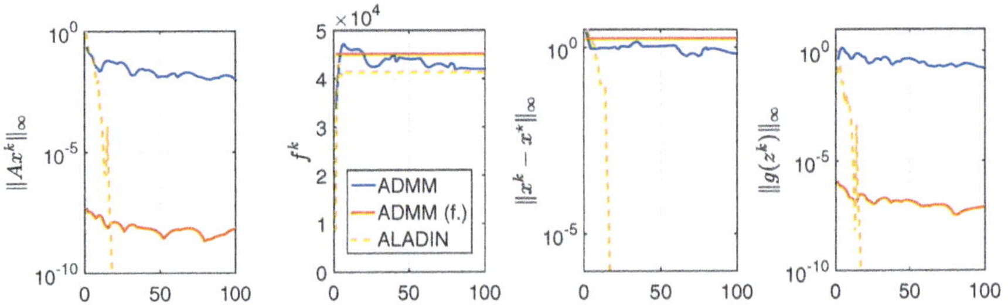

Fig. 3 Convergence behavior of ADMM with infeasible initialization, ADMM with feasible initialization (f.) for $\rho = 10^{12}$ and ALADIN

and the consensus step of ADMM enforce an alternating projection between the consensus constraint and the local nonlinear constraints. The progress in the nonlinear constraint violation $\|g(z^k)\|_\infty$ supports this statement. In its extreme, this behavior can be observed when using $\rho = 10^{12}$ depicted in Fig. 3. There, ADMM with feasible initialization produces very small consensus violations and nonlinear constraint violations at cost of almost no progress in terms of optimality.

Here the crucial observation is that in case of feasible initialization and large penalization parameter ρ ADMM produces *almost feasible* iterates at cost of slow progress in the objective function values. From this, one is tempted to conclude that also for infeasible initializations ADMM will likely converge, cf. Figs. 1, 2, and 3. This conclusion is supported by the rather small 57-bus test system. However, it deserves to be noted that this conclusion is in general not valid, cf. [1].

For ALADIN, we use $\rho = 10^6$ and $\mu = 10^7$. Comparing the results of ADMM with ALADIN, ALADIN shows superior quadratic convergence also in case of infeasible initialization. This is inline with the known convergence properties of ALADIN [9]. However, the fast convergence comes at the cost of increased communication overhead per step, cf. [6] for a more detailed discussion. Note that ALADIN involves a more complex coordination step, which is not straightforward to solve via neighborhood communication. Furthermore, tuning of ρ and μ can be difficult.

In power systems, usually feasibility is preferred over optimality to ensure a stable system operation. Hence, ADMM with feasible initialization and large ρ can in principle be used for OPF as in this case violation of power flow equations and generator bounds are expected to be small and one could at least expect certain progress towards an optimal solution. Following this reasoning several papers consider $\|A(x^k - z^k)\|_\infty < \epsilon$ as termination criterion [1, 4]. However, as shown in the example above, if ρ is large enough, and ADMM is initialized at a feasible point, this termination criterion can always be satisfied in just one iteration if ρ is chosen sufficiently large. Consequently, it is unclear how to ensure a certain degree of optimality. An additional question with respect to ADMM is how to obtain a feasible initialization. To compute such an initial guess, one has to solve a constrained nonlinear least squares problem solving the power flow equations subject to box constraints. This is itself a problem of almost the same complexity as the full OPF problem. Hence one would again require a computationally powerful central entity with full topology and parameter information. Arguably this jeopardizes the initial motivation for using distributed optimization methods.

4 Analysis of ADMM with Feasible Initialization for $\rho \to \infty$

The results above indicate that large penalization parameters ρ in combination with feasible initialization might lead to pre-mature convergence to a suboptimal solution. Arguably, this behavior of ADMM might be straight-forward to see from an optimization perspective. However, to the best of our knowledge a mathematically rigorous analysis, which is very relevant for OPF, is not available in the literature.

Proposition 1 (Feasibility and $\rho \to \infty$ imply $x_i^{k+1} - x_i^k \in \text{null}(A_i)$)
Consider the application of ADMM *(Algorithm* 1*) to Problem* (1)*. Suppose that, for all $k \in \mathbb{N}$, the local problems* (2) *have unique regular minimizers x_i^k.*[6] *For $\tilde{k} \in \mathbb{N}$, let $\lambda_i^{\tilde{k}}$ be bounded and, for all $i \in \mathcal{R}$, $z_i^{\tilde{k}} \in \mathcal{X}_i$. Then, the* ADMM *iterates satisfy*

$$\lim_{\rho \to \infty} x_i^k(\rho) - x_i^{\tilde{k}} \in \text{null}(A_i), \quad \forall k > \tilde{k}.$$

Proof The proof is divided into four steps. Steps 1–3 establish technical properties used to derive the above assertion in Step 4.

Step 1. At iteration $\tilde{k}$ the local steps of ADMM are

$$x_i^{\tilde{k}}(\rho) = \underset{x_i \in \mathcal{X}_i}{\text{argmin}}\ f_i(x_i) + \left(\lambda_i^{\tilde{k}}\right)^\top A_i x_i + \frac{\rho}{2}\left\| A_i\left(x_i - z_i^{\tilde{k}}\right)\right\|_2^2. \tag{6}$$

Now, by assumption all f_is are twice continuously differentiable (hence bounded on $\mathcal{X}_i$), $\lambda_i^{\tilde{k}}$ is bounded and all $z_i^{\tilde{k}} \in \mathcal{X}_i$. Thus, for all $i \in \mathcal{R}$, $\lim_{\rho \to \infty} x_i^{\tilde{k}}(\rho) = z_i^{\tilde{k}} + v_i^{\tilde{k}}$ with $v_i^{\tilde{k}} \in \text{null}(A_i)$.

[6] A minimizer is regular if the gradients of the active constraints are linear independent [14].

Step 2. The first-order stationarity condition of (6) can be written as

$$-\nabla f_i(x_i^{\tilde{k}}) - \gamma_i^{\tilde{k}\top} \nabla h_i(x_i^{\tilde{k}}) = A_i^\top \lambda_i^{\tilde{k}} + \rho A_i^\top A_i \left(x_i^{\tilde{k}} - z_i^{\tilde{k}}\right), \tag{7}$$

where $\gamma_i^{\tilde{k}\top}$ is the multiplier associated to h_i. Multiplying the multiplier update formula (3) with $A_i^\top$ from the left we obtain $A_i^\top \lambda_i^{k+1} = A_i^\top \lambda_i^k + \rho A_i^\top A_i(x_i^k - z_i^k)$. Combined with (7) this yields $A_i^\top \lambda_i^{\tilde{k}+1} = -\nabla f(x_i^{\tilde{k}}) - \gamma^{\tilde{k}\top} \nabla h_i(x_i^{\tilde{k}})$. By differentiability of f_i and h_i, compactness of $\mathcal{X}_i$ and regularity of $x_i^{\tilde{k}}$ this implies boundedness of $A_i^\top \lambda_i^{\tilde{k}+1}$.

Step 3. Next, we show by contradiction that $\Delta x_i^{\tilde{k}} \in \text{null}(A_i)$ for all $i \in \mathcal{R}$ and $\rho \to \infty$. Recall the coordination step (4b) in ADMM given by

$$\min_{\Delta x} \sum_{i \in \mathcal{R}} \frac{\rho}{2} \Delta x_i^\top A_i^\top A_i \Delta x_i + \lambda_i^{\tilde{k}+1\top} A_i \Delta x_i \quad \text{s.t.} \sum_{i \in \mathcal{R}} A_i(x_i^{\tilde{k}} + \Delta x_i) = 0. \tag{8}$$

Observe that any $\Delta x_i^{\tilde{k}} \in \text{null}(A_i)$ is a feasible point to (8) as $\sum_{i \in \mathcal{R}} A_i x_i^{\tilde{k}} = 0$. Consider a feasible candidate solution $\Delta x_i \notin \text{null}(A_i)$ for which $\sum_{i \in \mathcal{R}} A_i(x_i^{\tilde{k}} + \Delta x_i) = 0$. Clearly, $\lambda_i^{\tilde{k}+1\top} A_i \Delta x_i(\rho)$ will be bounded. Hence for a sufficiently large value of ρ, the objective of (8) will be positive. However, for any $\Delta x_i \in \text{null}(A_i)$ the objective of (8) is zero, which contradicts optimality of the candidate solution $\Delta x_i \notin \text{null}(A_i)$. Hence, choosing ρ sufficiently large ensures that any minimizer of (8) lies in $\text{null}(A_i)$.

Step 4. It remains to show $x_i^{\tilde{k}+1} = x_i^{\tilde{k}}$. In the last step of ADMM we have $z^{\tilde{k}+1} = x^{\tilde{k}} + \Delta x^{\tilde{k}}$. Given Steps (1)–(3) this yields $z^{\tilde{k}+1} = z^{\tilde{k}} + v^{\tilde{k}} + \Delta x^{\tilde{k}}$ and hence

$$\left\| A_i \left(x_i - z_i^{\tilde{k}+1}\right)\right\|_2^2 = \left\| A_i \left(x_i - z_i^{\tilde{k}} + v_i^{\tilde{k}} + \Delta x_i^{\tilde{k}}\right)\right\|_2^2 = \left\| A_i \left(x_i - z_i^{\tilde{k}}\right)\right\|_2^2.$$

Observe that this implies that, for $\rho \to \infty$, problem (6) does not change from step $\tilde{k}$ to $\tilde{k}+1$. This proves the assertion. ∎

Corollary 1 (Deterministic code, feasibility, $\rho \to \infty$ implies $x_i^{k+1} = x_i^k$) *Assuming that the local subproblems in* ADMM *are solved deterministically; i.e. same problem data yields the same solution. Then under the conditions of Proposition* 1 *and for* $\rho \to \infty$*, once* ADMM *generates a feasible point* $x_i^{\tilde{k}}$ *to Problem* (1)*, or whenever it is initialized at a feasible point, it will stay at this point for all subsequent* $k > \tilde{k}$.

The above corollary explains the behavior of ADMM for large ρ in combination with feasible initialization often used in power systems [1, 4]. Despite feasible iterates are desirable from a power systems point of view, the findings above imply that high values of ρ limit progress in terms of minimizing the objective.

Remark 1 (Behavior of ALADIN *for* $\rho \to \infty$*)* Note that for $\rho \to \infty$, ALADIN behaves different than ADMM. While the local problems in ALADIN behave similar to ADMM, the coordination step in ALADIN is equivalent to a sequential quadratic programming step. This helps avoiding premature convergence and it ensures decrease of f in the coordination step [9].

5 Conclusions

This method-oriented work investigated the interplay of penalization of consensus violation and feasible initialization in ADMM. We found that—despite often working reasonably with a *good choice* of ρ and infeasible initialization—in case of feasible initialization combined with large values of ρ ADMM typically stays feasible yet it may stall at a suboptimal solution. We provided analytical results supporting this observation. However, computing a feasible initialization is itself a problem of almost the same complexity as the full OPF problem; in some sense partially jeopardizing the advantages of distributed optimization methods. Thus distributed methods providing rigorous convergence guarantees while allowing for infeasible initialization are of interest. One such alternative method is ALADIN [9] exhibiting convergence properties at cost of an enlarged communication overhead and a more complex coordination step [6].

References

1. J. Guo, G. Hug, O.K. Tonguz, IEEE Trans. Power Syst. **32**(5), 3842 (2017). https://doi.org/10.1109/TPWRS.2016.2636811
2. D.K. Molzahn, F. Dörfler, H. Sandberg, S.H. Low, S. Chakrabarti, R. Baldick, J. Lavaei, IEEE Trans. Smart Grid **8**(6), 2941 (2017). https://doi.org/10.1109/TSG.2017.2720471
3. S. Boyd, N. Parikh, E. Chu, B. Peleato, J. Eckstein, Found Trends Mach. Learn. **3**(1), 1 (2011)
4. T. Erseghe, IEEE Trans. Power Syst. **29**(5), 2370 (2014). https://doi.org/10.1109/TPWRS.2014.2306495
5. B.H. Kim, R. Baldick, IEEE Trans. Power Syst. **15**(2), 599 (2000)
6. A. Engelmann, Y. Jiang, T. Mühlpfordt, B. Houska, T. Faulwasser, IEEE Trans. Power Syst. **34**(1), 584 (2019)
7. A. Engelmann, T. Mühlpfordt, Y. Jiang, B. Houska, T. Faulwasser, in *Proceedings of the American Control Conference (ACC)*, 2018, pp. 6188–6193. https://doi.org/10.23919/ACC.2018.8431090
8. A. Murray, A. Engelmann, V. Hagenmeyer, T. Faulwasser, IFAC-PapersOnLine **51**(28), 368 (2018). https://doi.org/10.1016/j.ifacol.2018.11.730. http://www.sciencedirect.com/science/article/pii/S2405896318334505. 10th IFAC Symposium on Control of Power and Energy Systems CPES 2018
9. B. Houska, J. Frasch, M. Diehl, SIAM J. Optim. **26**(2), 1101 (2016)
10. B.H. Kim, R. Baldick, IEEE Trans. Power Syst. **12**(2), 932 (1997)
11. D.P. Bertsekas, J.N. Tsitsiklis, *Parallel and Distributed Computation: Numerical Methods*, vol. 23 (Prentice Hall, Englewood Cliffs, 1989)

12. Y. Wang, W. Yin, J. Zeng, J. Sci. Comput. **78**(1), 29 (2019)
13. M. Hong, Z.Q. Luo, M. Razaviyayn, SIAM J. Optim. **26**(1), 337 (2016)
14. J. Nocedal, S. Wright, *Numerical Optimization* (Springer Science & Business Media, New York, 2006)

Security Analysis of Embedded HVDC in Transmission Grids

Marco Giuntoli and Susanne Schmitt

Abstract Security-constrained optimal power flow is widely used by grid operators during their short-term operations. Possible contingencies (e.g. outages of lines, generators) analyzed during this phase may be very critical for the stability of the power system. Thus, a simplified or reduced approach might not be desired. Moreover, the presence of embedded high-voltage direct current (HVDC) links (i.e. the DC link is done between two or more synchronous AC nodes) increases the degrees of freedom of the physical problem and thus an economically better working point for the system can be reached. To do this, a coordinated control between the AC and DC systems is required using one single stage of optimal power flow including the security assessment. In this work the SC-OPF (Security-Constrained Optimal Power Flow) approach is used to analyse the effects of contingency cases within the DC system.

Keywords Optimal power flow · Security-constrained optimal power flow · Embedded HVDC · N-1 criterion

1 Introduction

There is a world-wide trend towards increasing the share of renewable energy sources for electricity generation. Since the yield of renewable energy sources like e.g. solar and wind power is time-varying and depends on geographic location there is an increasing demand for large-scale power transfer over long distances. In recent years there have been more and more high-voltage direct current (HVDC) links for power transmission commissioned and more are yet to be planned and installed. In [1] an extensive overview of the advantages of DC technology for power transmission is given. For instance, in Germany there are several embedded HVDC

M. Giuntoli · S. Schmitt (✉)
ABB Corporate Research Center Germany, Ladenburg, Germany
e-mail: marco.giuntoli@de.abb.com; susanne.schmitt@de.abb.com

V. Bertsch et al. (eds.), *Advances in Energy System Optimization*,
Trends in Mathematics, https://doi.org/10.1007/978-3-030-32157-4_2

links (i.e. they connect synchronous AC nodes) planned in order to transfer wind power from the north to the high load regions in the south, see [2].

Optimal power flow is a widely discussed optimization problem to determine the optimal operation of controllable equipment in a power grid, see e.g. [3, 4]. For transmission grid operation the so-called $N - 1$ criterion as defined in [5] is a requirement. It means that for any single outage of a grid component (e.g. a generator, a line or a power transformer) the remaining grid must be able to operate safely within its limits. This leads to security-constrained optimal power flow problems (SC-OPF). For this class of problems an extensive overview is provided in [6]. SC-OPF problems can be formulated either in the preventive way, i.e. there is a single set of control variables that has to be feasible in all considered contingency cases or in the corrective way, i.e. for each contingency case there may be a different set of control variables, and their values can be reached by short-term control actions when a contingency case is taking place, see [7].

Optimal power flow formulations for hybrid AC/DC grids including security constraints are presented e.g. in [1, 8] and [9]. In [10] a corrective SC-OPF methodology is employed to investigate the capabilities of embedded HVDC systems to compensate for contingency cases in the AC system and thus to reduce redispatch by generators.

In this work we use SC-OPF to study the effects of the $N - 1$ criterion with emphasis on contingency cases within the embedded HVDC systems. To our knowledge the effects of $N - 1$ cases inside the DC systems have not yet been studied extensively in the literature. As in [10] we focus on the capabilities of the grid components and in particular the embedded HVDC systems to compensate for contingency cases and to avoid generator redispatch.

This work is structured as follows: in Sect. 2 we present the mathematical model for SC-OPF for hybrid AC/DC grids. Then, in Sect. 3 we describe the case study grid and scenario used as well as the implementation of the approach (Sect. 4). The case study results are presented and discussed in Sect. 5, and we give conclusions and suggestions for future work in the final Sect. 6.

2 The Mathematical Model

The mathematical model of the SC-OPF is used to perform a steady-state optimization for hybrid AC/DC grids, where the two types of grids are coupled via a set of power converters. The model is formulated such that just one single optimization problem can be used to solve the AC and DC grids simultaneously; this means in particular that no iterative algorithm is needed to reach the final solution.

The model is written to address one-phase electric systems, i.e. it can evaluate a mono-phase system or a multiphase system but in the latter case the system must be balanced and symmetric to consider only the positive sequence (Fortescue transformation).

All the variables and equations are expressed in complex polar form and they are differentiable and continuous. This means that discrete variables (e.g. limited number of tap changer positions or discontinuous active power range for the generators) are not considered. The set of variables are split into two parts: control (or independent) variables u representing power setpoints (active and reactive) for controllable devices as well as tap changer and phase changer positions for power transformers. State (or dependent) set of variables x consisting of voltage magnitude and phase at AC nodes, voltage at DC nodes as well as power flows over AC and DC branches.

Regarding the $N-1$ criterium, the model is able to take into account the outage of a branch (overhead line, power transformer or cable) or a generic equipment inside the system (power generator, power converter). Both the preventive and the corrective $N-1$ methodology can be used. Furthermore, there is no particular restriction about the grid topology. This means that it is possible to solve meshed AC and DC (e.g. multiterminal DC) synchronous and asynchronous grids.

The mathematical model can be formulated as the following nonlinear optimization problem:

$$\begin{aligned}
\underset{u_0,\ldots u_{N_c}}{\text{minimize}} \quad & p_0 f(u_0, x_0) + \sum_{c=1}^{N_c} p_c f(u_c, x_c) \\
\text{subject to} \quad & g_c(u_c, x_c) = 0 \;\; c = 0 \ldots N_c, \\
& h_c(u_c, x_c) \leq 0 \;\; c = 0 \ldots N_c, \\
& \underline{\Delta u_c} \leq u_c - u_0 \leq \overline{\Delta u_c},
\end{aligned} \tag{1}$$

where N_c is the number of contingency cases considered ($c = 0$ corresponds to the base case where no contingencies have happened). The set of equality constraints $g_c(u_c, x_c)$ describes the power balance for each node and equipment, while the set of inequality constraints $h_c(u_c, x_c)$ describes the physical limits of each component.

The deviation of the control variable in case of corrective $N-1$ method can be accordingly limited with the upper $\overline{\Delta u_c}$ and lower$\underline{\Delta u_c}$ bounds (e.g. active power setpoint deviation between the contingency cases and the base case).

The objective function f may be chosen as the generator dispatch cost in the system, system losses, voltage magnitude or angle variability, reactive power generated or another power system objective. It is also possible to provide flexible weights for the different contingency cases under consideration. In this work the objective function is defined as the economic dispatch cost in quadratic form:

$$\begin{aligned}
\zeta = & \sum_{g=1}^{N_g} \left\{ \alpha_g + \beta_g P_g + \gamma_g P_g^2 \right\} \\
& + \sum_{c=1}^{N_c} \left\{ \sum_{g=1}^{N_g} \left\{ c_g^+ \Delta p_{g,c}^+ + c_g^- \Delta p_{g,c}^- \right\} \right\}
\end{aligned} \tag{2}$$

Here N_g is the number of generators in the grid, P_g is the active power generated by the generator g, while α_g, β_g, γ_g are cost coefficients for the individual generators. $\Delta p^+_{g,c}$ and $\Delta p^-_{g,c}$ are the active power deviation (positive and negative) for each contingency and they are calculated with respect to the base case and weighed with the relative linear cost coefficients c^+_g and c^-_g.

The grid model is composed of the following items: nodes (AC and DC), power loads, branches and active equipments (generators, converters and reactive power sources like e.g. static VAR compensators (SVC)). Two power balance equations are given for the AC nodes and one for the DC nodes. In the subsequent paragraphs an overview over the models of the grid components is provided.

The following equation describes the active power balance for an AC node:

$$-p(v_{n,c}, \theta_{n,c}) + p_{g,c} + p_{h,ac,c} - p_{n,d} = 0, \tag{3}$$

where the subscripts n and c denote the n-th node and the c-th contingency case. The term $p(v_{n,c}, \theta_{n,c})$ is the active power flow into the n-th AC node where v and θ are the voltage magnitude and phase; the term $p_{g,c}$ is the power supplied by the g-th active source; the term $p_{h,ac,c}$ is the power supplied by the h-th converter; the term p_d is the power demand. In the same way it is possible to describe the reactive power balance for the AC nodes:

$$-q(v_{n,c}, \theta_{n,c}) + q_{g,c} + q_{h,ac,c} - q_{n,d} = 0. \tag{4}$$

For the DC nodes, only one power balance equation is needed:

$$-p(v_{n,c}) + p_{h,dc,c} = 0, \tag{5}$$

since in this case no power generators and loads are present. For both AC and DC nodes the voltage magnitude is constrained to its lower and upper limits.

The branches are modeled as generic passive components able to transfer current from one node to another according to Kirchhoff's circuit laws. The double bi-pole model with PI scheme has been used. This model can be applied to overhead lines, cables, power transformers (with and without tap changer and/or phase shifter) and to DC overhead lines and cables. Without loss of generality we assume the tap changer and phase shifter to be installed on the from-node of the branch and constrained with their limits. Both can be modelled with a pure gain δ and a complex rotation $e^{j\theta}$. For a generic branch b, contingency c and defining i as from-node and j as to-node, it is possible to describe the relationship between the current and voltage phasors as follows:

$$\begin{bmatrix} \dot{i}_i \\ \dot{i}_j \end{bmatrix} - \begin{bmatrix} \overline{y_{ii}}\delta^2_{b,c} & \overline{y_{ij}}\delta_{b,c}e^{j\theta_{b,c}} \\ \overline{y_{ji}}\delta_{b,c}e^{-j\theta_{b,c}} & \overline{y_{jj}} \end{bmatrix} \begin{bmatrix} \dot{v}_i \\ \dot{v}_j \end{bmatrix} = 0. \tag{6}$$

Exploiting the symmetry of the model, the matrix component with subscript ii is equal to the component with jj and, in the same way, the component with ij is equal to the component with ji. Those values arise from the longitudinal and transversal susceptance and conductance parameters. The equations that describe a double bi-pole model are used to model the active and reactive power flow through the branches, and thus those flows are used inside the nodes power balance equations (3), (4), (5). Obviously, the power flow (active power or apparent power or current module) can be constrained at both sides of the branch with an upper limit.

The power generators are generic active components able to inject active and/or reactive power into the system. The generator capability is defined as a domain created by the intersection of a rectangle (active and reactive upper and lower bounds) and a circle (apparent power bound). Additional linear bounds between active and reactive power can be defined in order to customize the capability domain.

In a similar way, by neglecting the active power, voltage controlling reactive power sources can be used inside the model: in this case, a linear relationship between voltage magnitude and reactive power output is used.

The power converters are able to transfer power in both directions between AC and DC nodes. The converter model takes into account the power balance between the AC and DC ports with the losses due to its power electronics. Moreover, the converters may have capability domains described in the same way as for generators. The converter power balance equation is written as follows:

$$p_{h,ac,c} + p_{h,dc,c} + p_{h,loss,c} = 0, \tag{7}$$

where the subscripts $loss$ denotes the power losses inside the h-th converter and the c-th contingency. The AC side current is determined by the following equation that describes the power balance in the AC converter side:

$$\frac{\sqrt{p_{h,ac,c}^2 + q_{h,c}^2}}{\sqrt{3}\, v_{h,ac,c}} - i_{h,ac,c} = 0. \tag{8}$$

Finally, the power losses are calculated by a formula that creates a quadratic relationship between the current $i_{h,ac,c}$ and losses $p_{h,loss,c}$ (see e.g. [11] and [12]) with parameters α_h, β_h and γ_h that may be individual for each converter:

$$\alpha_h + \beta_h i_{h,ac,c} + \gamma_h i_{h,ac,c}^2 - p_{h,loss,c} = 0. \tag{9}$$

This model is suited for steady-state analyses, so that transient behaviour, e.g. fault and stability analysis cannot be investigated with it.

3 Case Study

The case study in this work is based on the IEEE118 test grid (see [13]).

The generator costs have been modified in order to encourage power transfer from the (low cost) western zone to the (high cost) eastern zone (see Fig. 1). In particular, the generation cost in the western zone have been set to zero to emulate renewable generation, while the cost coefficients in the western zone have not been modified.

The focus of our analysis lies on one additional power corridor between the nodes 8 and 65 (highlighted in red in Fig. 1). There are three options for this considered corridor: an AC corridor, a monopolar HVDC corridor as well as a bipolar HVDC corridor with identical total power ratings of 500 MVA and total length of 500 km. The electrical parameters of the AC corridor as well as the resistance of the DC lines are taken from [14]. Due to the huge length of the corridor the AC line has been split in three line segments. The HVDC converters exhibit a loss model with a quadratic relationship to the AC current (see [15]).

For all three configurations the effects of security constraints on the corridor are analysed. For both the AC corridor and the monopolar HVDC corridor the considered contingency implies the outage of the whole corridor, while for the bipolar configuration an $N-1$ contingency case involved the outage of one of the two poles.

In order to discourage redispatch of generators in the contingency case, we require the generator active and reactive power output to be identical in the base case and the contingency case, while HVDC converter setpoints may differ between base case and contingency case. This means that the generators follow a preventive security strategy while HVDC converters follow a corrective security strategy. The same behaviour can be achieved by imposing very high redispatch cost on the generators. The possibly changing losses after an outage are balanced by adapted voltage profiles. In addition, the setpoints for transformers are fixed.

4 Implementation

A software tool for solving the security-constrained optimal power flow problem as described in Sect. 2 has been developed in C++ using Microsoft® Visual Studio™. Inside this software the open-source library IPOPT (see [16]) is used as solver for non-linear programming based on an interior point method. To speed up the optimization process and to increase accuracy, we derived the analytic forms of Jacobian and Hessian matrices of the objective function and the constraints and pass them directly to the solver instead of using numerical differentiation.

The input data as well as the optimization results including all control and state variables are given as text files. In the results discussion below (Sect. 5) we summarize these results in the form of the key performance indicators dispatch

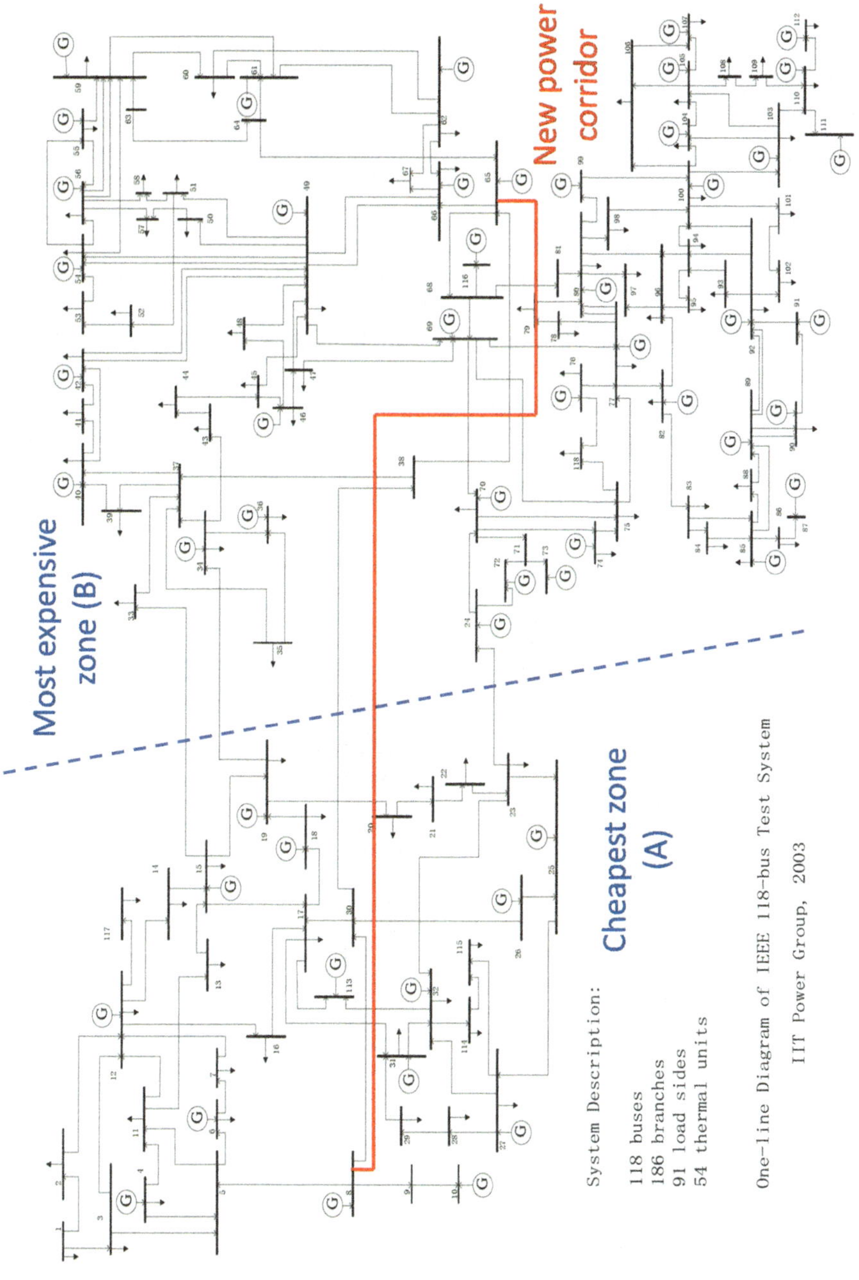

Fig. 1 Illustration if the IEEE118 test grid, the two generator cost zones and the additional corridor under investigation [17]

cost, power flow over the considered corridor, active power losses and voltage angle deviation.

5 Results

The result of the SC-OPF is analysed in terms of dispatch cost (Fig. 2), power flow over the considered corridor (Fig. 3), active power losses in the system (Fig. 4) and voltage angle difference along the corridor (Fig. 5).

The first observation is that in the cases without the $N-1$ criterion, all the values for the monopolar and the bipolar HVDC corridors coincide, which is obvious since they have identical parameters and rating. When looking at the dispatch cost in Fig. 2 it can be observed that the cost is lowest for both the HVDC variants without $N-1$ criterion. With $N-1$ criterion both HVDC variants have lower dispatch cost than the AC variant, while the bipolar configuration has the lowest dispatch cost. This can be explained by the higher efficiency of the HVDC systems compared to the AC corridor (lower losses over long distances) and the higher redundancy of the bipolar configuration compared to the monopolar one. This way more power can

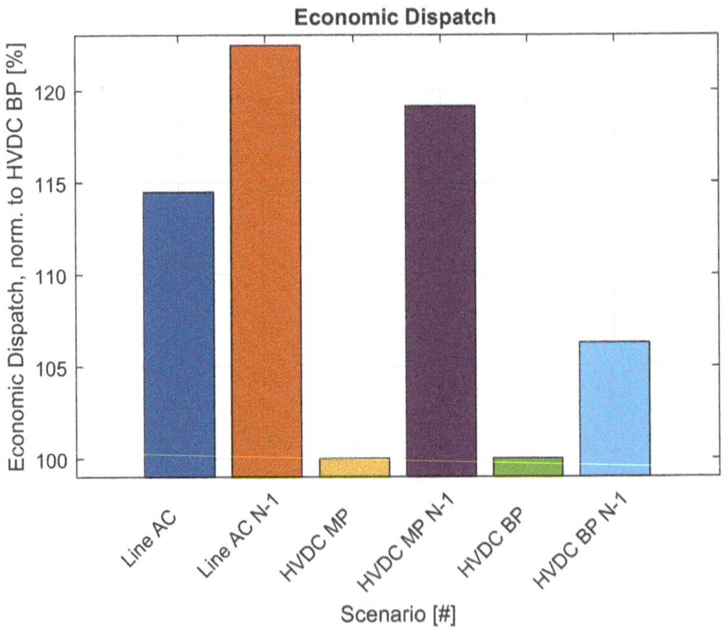

Fig. 2 Dispatch cost relative to bipolar HVDC scenario without $N-1$. Red, dark blue: AC line with/without $N-1$; violet, yellow: monopolar HVDC with/without $N-1$; light blue, green: bipolar HVDC with/without $N-1$

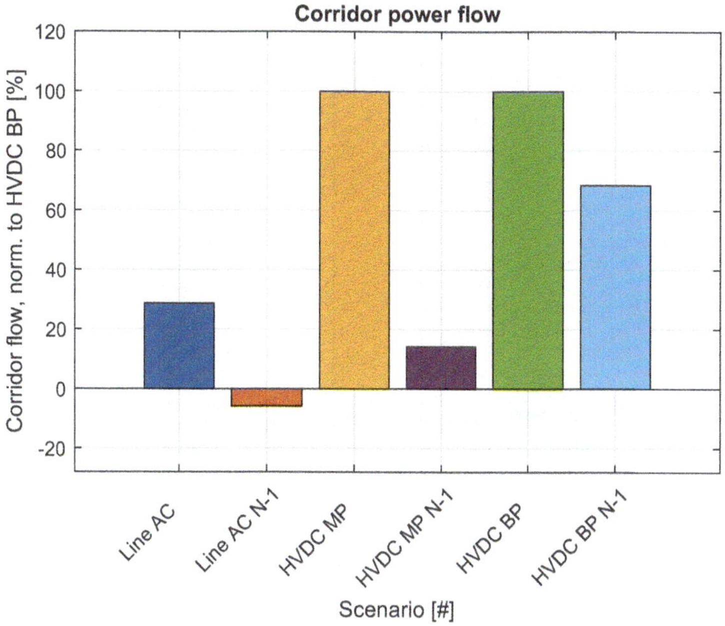

Fig. 3 Power flow over corridor relative to bipolar HVDC scenario without $N-1$ (457 MW)

be transferred from the low cost generation zone to the high cost generation zone which enables a generation dispatch at lower total cost.

This explanation is confirmed by the power flow over the corridor (Fig. 3), where the bipolar HVDC configuration is able to transfer the most power from the low cost generation zone to the high cost generation zone. The monopolar HVDC corridor is transferring significantly less power in the $N-1$ case since it has to prepare for the outage of the full corridor. The AC corridor is not even able to maintain the power flow direction in the $N-1$ case. The significant deviations in the power flows between the AC and monopolar HVDC configuration can be explained by voltage limits and differences in the generator dispatch.

The active power losses in the system can be seen in Fig. 4. There we did not observe a clear pattern, which can be explained by the fact that in all the scenarios the generator dispatch and thus the power flows through the whole system are very different, and the losses were not the objective of the optimization performed.

When looking at the angle deviations in Fig. 5 is can be observed that the AC corridor exhibits a large angle deviation in the case without security criterion, while all the HVDC configurations are able to maintain smaller angle deviations which indicates a higher system stability.

In Table 1 the voltage magnitudes and reactive power injected at the corridor terminal nodes 8 and 65 are listed for the $N-1$ scenario before and after the contingencies. It can be observed that the bipolar HVDC configuration is able to

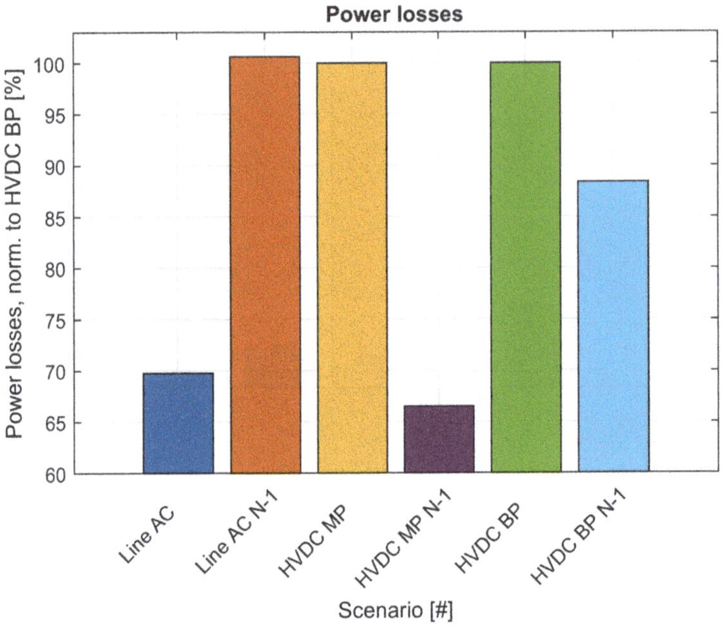

Fig. 4 Active power losses in the system relative to bipolar HVDC scenario without $N - 1$ (120 MW)

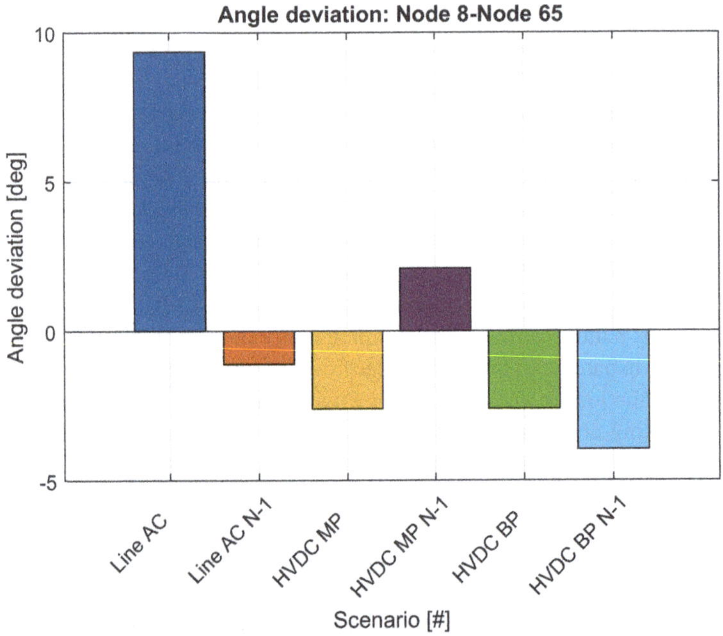

Fig. 5 Angle Deviation over corridor in degrees

Table 1 Voltage magnitude and reactive power injection at nodes 8 and 65 for all three corridor configurations in the base case and the contingency case of the $N-1$ scenarios

Case	V(8) [p.u.]	V(65) [p.u.]	Q(8) [MVAr]	Q(65) [MVAr]
AC; base	0.980	0.968	98.2	118.6
AC, cont.	1.046	0.965	0.0	0.0
DC MP; base	1.039	1.012	−19.0	−96.1
DC MP, cont.	0.973	1.021	0.0	0.0
DC BP; base	1.004	1.023	−58.7	−5.1
DC BP, cont.	0.998	1.023	−56.5	0.0

maintain a stable voltage level close to the nominal voltage by providing reactive power to the grid. In contrast, for the AC configuration the voltage magnitude at node 8 changes significantly because of high generation in the region. The monopolar HVDC configuration is also stabilizing the voltage by reactive power injection but in the outage case there is the same effect as for the AC corridor.

6 Conclusion and Future Work

A mathematical model and a software for security-constrained optimal power flow for generic hybrid AC/DC grids has been developed. In this work it has been used to study the effects of the $N-1$ criterion on embedded HVDC systems.

The above results show that for the bipolar HVDC configuration the power plant dispatch with lowest cost could be established when the $N-1$ criterion is employed. This study has been performed on the IEEE118 grid with a single corridor under consideration. The results indicate the benefits of employing bipolar HVDC systems in order to increase the social welfare.

In order to draw conclusions for real transmission systems, additional studies on real grid configurations with possibly several embedded HVDC links would have to be performed.

Additionally, an investigation including contingency cases both in the AC and DC parts of the power system would be interesting to investigate the capabilities of embedded HVDC to compensate for general $N-1$ cases and thus to reduce redispatch cost for grid operators.

Acknowledgements This research was partly supported by the Federal Ministry of Education and Research (BMBF) within the framework of the project "Neue EnergieNetzStruktURen für die Energiewende"İ ENSURE (FKZ 03SFK1N0).

References

1. R. Wiget, *Combined AC and Multi-terminal HVDC Grids – Optimal Power Flow Formulations and Dynamic Control*, Dissertation thesis, ETH Zurich, 2015
2. Bundesnetzagentur, *Bestätigung Netzentwicklungsplan Strom*, 2017
3. H. Dommel, W. Tinney, Optimal power flow solutions. IEEE Trans. Power Syst. **PAS-87**(10), 1866–1876 (1968)
4. O. Alsac, B. Stott, Optimal load flow with steady-state security. IEEE Trans. Power Syst. **PAS-93**(3), 745–751 (1974)
5. Network Code on Operational Security, Entso-E Tech. Rep., Sept 2013
6. F. Capitanescu, J.L. Martinez Ramos, P. Panciatici, D. Kirschen, A. Marano Marcolini, L. Platbrood, L. Wehenkel, State-of the-art, challenges, and future trends in security constrained optimal power flow. Electr. Power Syst. Res. **81**(8), 1731–1741 (2011)
7. G. Papaefthymiou et al., Methods for optimization of power transits": toolbox for Common Forecasting, Risk assessment, and Operational Optimization in Grid Security Cooperations of Transmission System Operators (TSOs): toolbox for Common Forecasting, Risk assessment, and Operational Optimization in Grid Security Cooperations of Transmission System Operators (TSOs), 2015, Available: https://www.research-collection.ethz.ch/handle/20.500.11850/114121. Accessed on 11 Dec 2018
8. S. Chatzivasileiadis, *Power System Planning and Operation Methods Integrating the Controllability of HVDC*, Dissertation thesis, ETH Zurich, 2013
9. R. Wiget, G. Andersson, *Optimal Power Flow for Combined AC and Multi-Terminal HVDC Grids based on VSC Converters*, IEEE Power Energy Society (PES) General Meeting, San Diego, 22–26 July 2012
10. N. Meyer-Huebner, S. Weck,F. Bennewitz, M. Suriyah, K. Bhalodi, M. Giuntoli, V. Biagini, A. Krontiris, A. Wasserab, M. Ndreko, M. Wiest, T. Leibfried, J. Hanson, *N-1-Secure Dispatch Strategies of Embedded HVDC Using Optimal Power Flow*, IEEE Power Energy Society (PES) General Meeting, Portland, 5–10, Aug, 2018, pp. 1–8
11. D. Var Hertem, O. Gomis-Bellmnt, H. Liang, *HVDC Grids for Offshore and Supergrids of the Future* (IEEE Press/Wiley, New Jersey, 2016), pp. 300–301
12. G. Daelemans, K. Srivastava, M. Reza, S. Cole, R. Belmans, *Minimization of steady-state losses in meshed networks using VSC HVDC*, IEEE Power Energy Society (PES) General Meeting, Alberta, 26–30 July 2009, pp. 1–5
13. www.ee.washington.edu, IEEE 118 test grid, Available: http://www.ee.washington.edu/research/pstca/. Accessed on 05 May 2018
14. A.P. Melipoulos, *Power System Modeling Analysis and Control*, chapter 6 (Marcel Dekker Inc, Atlanta, 2004)
15. D. Var Hertem, O. Gomis-Bellmnt, H. Liang, *HVDC Grids for offshore and supergrids of the future* (IEEE Press/Wiley, New Jersey, 2016), pp. 300–301
16. A. Wächter, L.T. Biegler, On the implementation of an interior-point filter line-search algorithm for large-scale nonlinear programming. Math. Program. **106**(1), 25–57 (2006)
17. A.R. Al-Roomi, *Power Flow Test Systems Repository* (Electrical and Computer Engineering, Dalhousie University, Halifax, 2015). https://al-roomi.org/power-flow

Multi-area Coordination of Security-Constrained Dynamic Optimal Power Flow in AC-DC Grids with Energy Storage

Nico Huebner, Nils Schween, Michael Suriyah, Vincent Heuveline, and Thomas Leibfried

Abstract In times of growing integrated electricity markets and needed coordination of large inter-regional physical power flows, multi-area Optimal Power Flow (OPF), also referred to as distributed OPF, has gained importance in research. However, the conventional OPF is only of limited use since a TSO is strongly interested in N-1 security. Furthermore, time-dependent constraints such as generator ramping or energy storage limits play a growing role. Consequently, a Security-Constrained Dynamic OPF (SC-D-OPF) includes both N-1 security and quasi-stationary dynamics. We present a decoupling approach to compute an SC-D-OPF by coordination among inter-connected areas. Privacy is maintained by implementing an Alternating Direction of Multipliers Method (ADMM), where only results of boundary variables are exchanged with a neighbor. We show the functionality of the approach in a small test case, where the distributed result is close to that of a centralized optimization.

Keywords Multi-area optimal power flow · Distributed optimal power flow · N-1 security · Energy storage · Dynamic optimal power flow · Multi-period optimal power flow · Embedded HVDC · AC-DC grids

---

N. Huebner (✉) · M. Suriyah · T. Leibfried
Institute of Electric Energy Systems and High-Voltage Technology (IEH), Karlsruhe Institute of Technology (KIT), Karlsruhe, Germany
e-mail: nico.meyer-huebner@kit.edu

N. Schween · V. Heuveline
Engineering Mathematics and Computing Lab (EMCL), Interdisciplinary Center for Scientific Computing (IWR), Heidelberg, Germany

V. Bertsch et al. (eds.), *Advances in Energy System Optimization*,
Trends in Mathematics, https://doi.org/10.1007/978-3-030-32157-4_3

1 Introduction

The increasing penetration of Renewable Energy Sources (RES) leads to a power system operation closer to its operational limits. Enhanced methods and tools will be crucial for a secure, but also efficient and cheap planning and operation of the power grid. A Transmission System Operator (TSO) will have larger assets of controllable devices at hand, such as Voltage Source Converter (VSC-) based HVDC-systems and energy storage systems. Combined with the need to assure N-1 security, this requires a Security-Constrained Dynamic Optimal Power Flow (SC-D-OPF), which incorporates both N-1 security and multiple time steps into the optimization. First research in that area was done in [1], where not only power but also necessary energy reserves can be determined to ensure N-1 security. In [2], successive linear algorithms and approximated power flows are used to solve large-scale SC-D-OPF problems in a European context. The authors of [3] propose an SC-D-OPF model including uncertainty of wind power or equipment availability, which is solved in a two-stage stochastic program. In [4], SC-D-OPF is used as inner iteration in a hierarchical approach to optimize a central deployment signal which is sent to the units able to provide a reserve. Furthermore, [5–7] include energy storage systems in an SC-D-OPF calculation. A further extension is shown in [8], where SC-D-OPF is solved in a hybrid AC-DC grid with energy storage.

Due to the increasing complexity of the power system and an operation closer to network limitations, central coordination in large scale networks comes with major computational burdens. Furthermore, privacy can be a concern for each transmission system operator (TSO) controlling a certain region. Subsequently, the interest in distributed optimization, also referred to as multi-area optimization has substantially grown in recent years. An early overview of distributed OPF algorithms can be found in [9] and the most recent developments are examined in detail in [10]. The Alternating Direction of Multipliers Method (ADMM) [11], has become a popular branch to tackle the non-convex AC OPF problem [12–14]. Each area uses variable duplicates from neighboring areas and is solved to optimality. Penalty terms, which are calculated from the coupling variable deviation and Lagrangian multipliers, are added to the objective function to push two neighbors towards a common boundary solution. ADMM was applied to a hybrid AC-DC grid in [15]. Multi-area optimization is applied to D-OPF with storage in [16] using OCD, but to the best of our knowledge, no attempts have been made toward distributing SC-D-OPF.

2 Security-Constrained Dynamic Optimal Power Flow (SC-D-OPF)

The optimization horizon is defined by a set of time steps $\mathcal{T} = \{1, \dots, T\}$ with T the total number of considered time steps. Furthermore, we define a scenario set $\mathcal{C} = \{0, \dots, C\}$. Here, scenario $c = 0$ defines the base case, i.e. the original fault-free network. A total of C possible contingencies is added by including modified versions of the base case to the scenario set. A modification could be the outage of an AC line, a DC line or a converter. With $x^{t,c}$ the optimization variables of time step t and scenario c, the full set of optimization variables is collected in

$$x = \left((x^{1,0})^\top, \ldots, (x^{1,C})^\top, \ldots, (x^{T,0})^\top, \ldots, (x^{T,C})^\top\right)^\top. \tag{1}$$

The full problem has the form (2):

$$\underset{x}{\text{minimize}} \quad \sum_{t \in \mathcal{T}} \sum_{c \in \mathcal{C}} p^{t,c} \cdot f(x^{t,c}) \tag{2a}$$

$$\text{subject to} \quad h_{\text{base}}(x^{t,c}) \leq 0 \quad \forall t \in \mathcal{T}, c \in \mathcal{C} \tag{2b}$$

$$h_{\text{dyn}}(x^{t,0}, x^{t-1,0}) \leq 0 \quad \forall t \in \mathcal{T} \setminus 1 \tag{2c}$$

$$h_{\text{N-1}}(x^{t,0}, x^{t,c}) \leq 0 \quad \forall t \in \mathcal{T}, c \in \mathcal{C} \setminus 0. \tag{2d}$$

It minimizes the sum of weighted costs f over all scenarios. Factor $p^{t,c}$ defines a probability of each contingency to enable a risk-based OPF. Constraints h_{base} describe the basic constraints of a conventional OPF, which must be fulfilled during each scenario (t, c). That includes AC and DC node power balance; thermal AC and DC branch limits; AC and DC voltage limits; quadratic loss function of AC-DC converters; and operational limits of generators, converters or sheddable loads. Time-dependent constraints, such as storage energy limits or generator ramping limits, are collected in h_{dyn}. Those constraints only apply to the base case ($c = 0$), since set points after an outage are coupled solely to the respective base case of the identical time step. This N-1 coupling is modeled by the constraints $h_{\text{N-1}}$ and guarantees N-1 security. Here, the coupling between each contingency and the base case is explicitly modeled. That is, set point deviation limits from *before* to *after* the occurrence of an outage can be defined. Note that for the sake of readability, we omit equality constraints which can be interpreted as reformulated inequalities as well. We refer to [8] for more details on the modeling.

3 Distributed Formulation

For the sake of readability, we write (2) in the form (3):

$$\underset{x}{\text{minimize}} \quad f(x) \tag{3a}$$

$$\text{subject to} \quad h(x) \leq 0. \tag{3b}$$

Let R non-overlapping regions be defined in $\mathcal{R} = \{1, \ldots, R\}$, that is, each node belongs to exactly one region. An equivalent problem as (3), but in separable form, can be written as (4):

$$\underset{x_k, \forall k \in \mathcal{R}}{\text{minimize}} \quad \sum_{k \in \mathcal{R}} f_k(x_k) \tag{4a}$$

$$\text{subject to} \quad h_k(x_k) \leq 0, \quad \forall k \in \mathcal{R} \tag{4b}$$

$$\sum_{k \in \mathcal{R}} A_k x_k = 0. \tag{4c}$$

Here, $x_k \in \mathbb{R}^{l_k}$, $h_k : \mathbb{R}^{l_k} \rightarrow \mathbb{R}^{m_k}$ and $f_k : \mathbb{R}^{l_k} \rightarrow \mathbb{R}$ are optimization variables, non-linear constraints and objective function, respectively, in a region k. Matrix $A_k \in \mathbb{R}^{n \times l_k}$ maps x_k onto the full set of n coupling constraints and enforces consensus between regions. Thus, (4b) forms independent local SC-D-OPF constraints and (4c) form the coupling constraints which recuperate a feasible overall power flow.

4 Network Decomposition in AC-DC Grids

To allow for a separable formulation, the network is first partitioned and then decomposed.

Network partitioning can be crucial for distributed algorithms to achieve good performance. We believe that network partitions are inherently given by structural responsibilities. For example, a region or control area could represent one TSO, multiple TSOs or a whole country. Thus, the number and dimension of AC regions are historically known. However, responsibilities in overlaying DC networks are yet to be defined and various options are thinkable [15]. We choose a *Shared-DC* approach. Here, we define $R = R^{\text{AC}}$ hybrid AC-DC regions, where each AC region may also contain DC nodes. Thus, each existing TSO gains control over converters in its own control area. Let $\mathcal{N}_k$ identify all nodes in region k. Herewith, $\mathcal{N}_k^{\text{AC}}$ collects all AC nodes and $\mathcal{N}_k^{\text{DC}}$ all DC nodes in region k.

4.1 Decoupling of AC or DC Tie Line

In principle, the decoupling of AC and DC tie line is identical, except for extended consensus constraints due to complex voltage and power in the AC case (Fig. 1).

The original line is cut into two halves; auxiliary nodes (m, n) and auxiliary generators (a, b) are added at both open ends. Thus, node sets are augmented to $\mathcal{N}_A \leftarrow \{\mathcal{N}_A, m\}$, $\mathcal{N}_B \leftarrow \{\mathcal{N}_B, n\}$. To guarantee a feasible power flow, the voltage

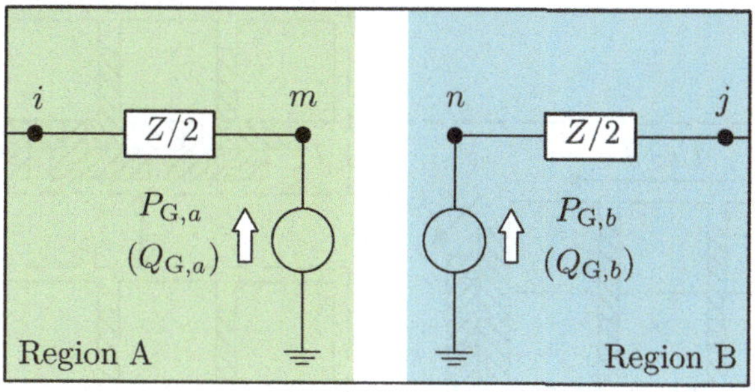

Fig. 1 Decoupling model of a tie line (AC or DC) between nodes i and j. Two auxiliary nodes (m and n) and two auxiliary generators (a and b) are added at the middle. Reactive power source Q_G is only added for an AC tie line

must be equal at nodes m and n. Furthermore, the generators must produce the same amount of power of the opposite sign. In the case of an AC tie line ($i \in \mathcal{N}_A^{AC}$, $j \in \mathcal{N}_B^{AC}$), this leads to boundary conditions including complex voltage and both active and reactive power. For time step and scenario (t, c), we have:

$$|V_{AC,m}^{t,c}| = |V_{AC,n}^{t,c}| \tag{5a}$$

$$\angle V_{AC,m}^{t,c} = \angle V_{AC,n}^{t,c} \tag{5b}$$

$$P_{G,a}^{t,c} = -P_{G,b}^{t,c} \tag{5c}$$

$$Q_{G,a}^{t,c} = -Q_{G,b}^{t,c}. \tag{5d}$$

In the case of a DC tie line ($i \in \mathcal{N}_A^{DC}$, $j \in \mathcal{N}_B^{DC}$), only real voltage and active power must meet the constraints. For time step and scenario (t, c), we have:

$$V_{DC,m}^{t,c} = V_{DC,n}^{t,c} \tag{6a}$$

$$P_{G,a}^{t,c} = -P_{G,b}^{t,c}. \tag{6b}$$

Equations (5) and (6) are the only boundary constraints.

4.2 Building Consensus Constraint Matrix A

Since $x_k^{t,c}$ uses the augmented node sets of region k, the auxiliary generator variables are included. Thus, consensus constraints (5)–(6) between region A and B can be written to

$$\tilde{A}_A^{t,c} x_A^{t,c} + \tilde{A}_B^{t,c} x_B^{t,c} = 0. \tag{7}$$

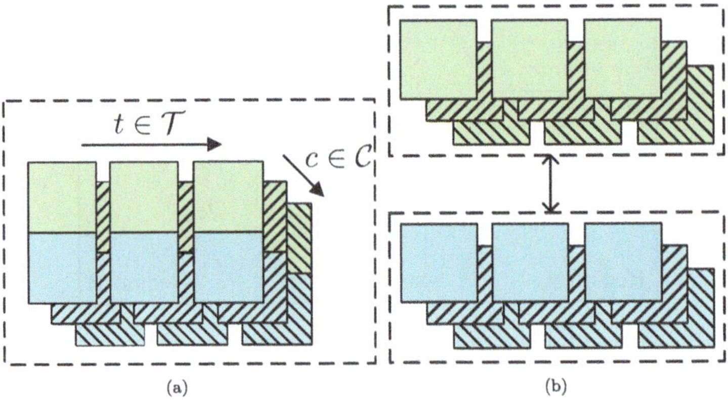

Fig. 2 Schematic problem decomposition for two areas (blue and green) under consideration of three time steps and two contingencies. A dashed box represents an optimization problem. (**a**) Centralized SC-D-OPF. (**b**) Distributed SC-D-OPF

This must be fulfilled for every considered time or contingency scenario (Fig. 2). Then we have

$$\begin{bmatrix} \tilde{A}_A^{1,0} & & & \tilde{A}_B^{1,0} & & \\ & \ddots & & & \ddots & \\ & & \tilde{A}_A^{T,C} & & & \tilde{A}_B^{T,C} \end{bmatrix} \begin{bmatrix} x_{\mathrm{A}}^{1,0} \\ \vdots \\ x_{\mathrm{A}}^{T,C} \\ x_{\mathrm{B}}^{1,0} \\ \vdots \\ x_{\mathrm{B}}^{T,C} \end{bmatrix} = 0. \tag{8}$$

In a more compact form, (8) can be written as

$$\sum_{k \in \{A,B\}} A_k x_k = 0. \tag{9}$$

5 Implemented ADMM Algorithm

The general idea of ADMM, see also [11], is the following. Augmented regional OPFs are solved and the deviation of boundary variables from a fixed auxiliary variable z, which is information stemming from neighboring regions, is penalized. The regions then exchange information and z is updated. The update is re-distributed to the local agents (areas) for a new OPF calculation until consensus between regions is achieved. It is constructed an augmented Lagrangian of the form

$$\mathcal{L}(x, z, \lambda) = \sum_{k \in \mathcal{R}} \left\{ f_k(x_k) + \lambda_k^\top A_k(x_k - z_k) \right.$$
$$\left. + \frac{\rho}{2} ||A_k(x_k - z_k)||_W^2 \right\}, \tag{10}$$

with penalty parameter $\rho \in \mathbb{R}$. Dual variables of the consensus constraints are denoted with $\lambda_k \in \mathbb{R}^{n \times 1}$, and $W \in \mathbb{R}^{n \times n}$ is a positive definite, diagonal weighting matrix,[1] where each entry is related to one coupling constraint. We refer to $W(S)$ if the entry is related to a power variable, and to $W(V)$ if the entry is related to a voltage variable. In ADMM literature, W is the identity matrix. The main steps during one iteration of the solving process are

$$1.\ x = \operatorname{argmin}_{x \in \mathcal{X}}\ \mathcal{L}(x, z, \lambda) \tag{11a}$$

$$2.\ z = \operatorname{argmin}_{z \in \mathcal{Z}}\ \mathcal{L}(x, z, \lambda) \tag{11b}$$

$$3.\ \lambda \leftarrow \lambda + \rho W A(x - z) \tag{11c}$$

with $z = [z_1^\top \ \dots \ z_R^\top]^\top$. The first step (11a) minimizes a non-linear problem, where constraint region

$$\mathcal{X} = \{x | h_k(x_k) \leq 0, \forall k \in \mathcal{R}\} \tag{12}$$

enforces local constraints (4b). Since z is fixed, (11a) is in fact a series of R independent problems which can be calculated in parallel. The second step minimizes a coupled quadratic problem, where

$$\mathcal{Z} = \left\{ z | \sum_{k \in \mathcal{R}} A_k z_k = 0 \right\} \tag{13}$$

enforces consensus of auxiliary variables z. In fact, (11b) calculates the average value between two consensus variables of neighboring regions [11, 13] and the problem can be reduced to

$$z = \operatorname{argmin}_{z \in \mathcal{Z}} \sum_{k \in \mathcal{R}} \frac{\rho}{2} ||A_k(x_k - z_k)||_W^2. \tag{14}$$

[1] A weighted norm is calculated with $||X||_W^2 = X^\top W X$.

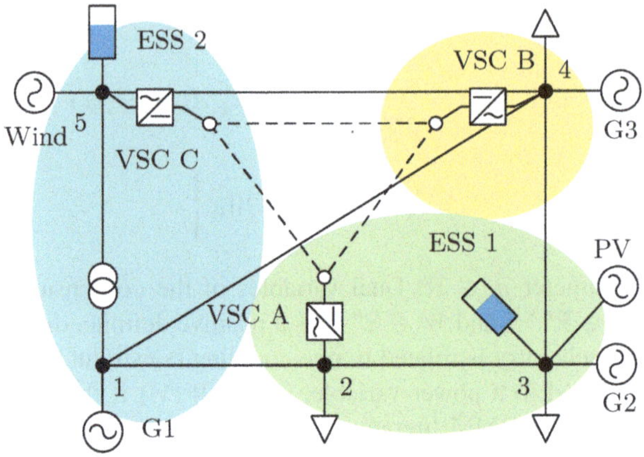

Fig. 3 Test system with embedded HVDC, energy storage and RES. Partitioning into 3 control areas

The weighting factors ρ and W can be neglected in (14), if the same weight is assigned to two coupled variables, which is a reasonable choice. Once a region has gathered necessary neighbor information, this step can be calculated locally as well [12]. In the third step (11c), dual variables are updated based on a weighted distance between x and z. Again, with given local x_k and z_k, each region can calculate λ_k independently (Fig. 3).

In ADMM, an update rule on ρ can be useful to enforce consensus. A $\rho_k \in \mathbb{R}$ is assigned to each region, which can be increased depending on the local residual, see (18b). If the residual has not decreased sufficiently compared to the previous iteration (indicator $0 < \Theta \in \mathbb{R} < 1$), the penalty is increased by a constant factor of $\tau \in \mathbb{R} > 1$. The penalty parameter must be chosen carefully since it is widely known to be crucial for good convergence behavior [17]. An overview of the implemented ADMM is given in Algorithm 1.

6 Results

6.1 Test Scenario

We use the illustrative AC-DC test system from [15], see Tables 1 and 2 for line and generator parameters, respectively. The generator at *Node 5* is interpreted as a wind park and a PV park is connected to *Node 3*. RES upper power limit and load are variable over time. Forecast power profiles are taken from the Belgian transmission system operator *Elia*, which provides detailed RES generation and load data on its website. Adapted profiles for usage in the 5-bus test system are shown in Figs. 6 and 7. We assume quadratic generator cost functions as in [18] (see Table 3) and one

Algorithm 1 ADMM

1: **Initialization:** Weighting matrix W, tolerance ϵ; for all $k \in \mathcal{R}$: initial guesses z_k, penalty parameters $\rho_k = \rho$, dual variables $\lambda_k = \mathbf{0}$, local solutions $x_k = \infty$, local residues $\Gamma_k = \infty$.
2: **while** $||Ax||_\infty > \epsilon$ and $||x - z||_\infty > \epsilon$ **do**
3: Solve for all $k \in \mathcal{R}$ the decoupled NLPs

$$\underset{x_k}{\text{minimize}} \;\; f_k(x_k) + \lambda_k^\top A_k x_k + \frac{\rho_k}{2} ||A_k(x_k - z_k)||_W^2 \tag{15a}$$

$$\text{subject to} \;\; h_k(x_k) \leq 0 \tag{15b}$$

4: Solve the coupled averaging step

$$\underset{z}{\text{minimize}} \;\; \sum_{k \in \mathcal{R}} ||A_k(x_k - z_k)||_2^2 \tag{16a}$$

$$\text{subject to} \;\; \sum_{k \in \mathcal{R}} A_k z_k = 0 \tag{16b}$$

5: Update dual variables for all $k \in \mathcal{R}$

$$\lambda_k \leftarrow \lambda_k + \rho_k W A_k(x_k - z_k) \tag{17}$$

6: Calculate local residues and penalty parameter updates for all $k \in \mathcal{R}$

$$\Gamma_k^+ = ||A_k(x_k - z_k)||_\infty \tag{18a}$$

$$\rho_k \leftarrow \begin{cases} \rho_k & \text{if } \Gamma_k^+ \leq \Theta \Gamma_k \\ \tau \rho_k & \text{otherwise} \end{cases} \tag{18b}$$

7: Update $\Gamma_k \leftarrow \Gamma_k^+$ for all $k \in \mathcal{R}$
8: **end while**

cost coefficient for all reactive power injections. RES power injection is prioritized by assigning a low fuel price. Two energy storage systems with characteristics from Table 4 are connected to *Node 3* and *5*, respectively. Line flow limits of 400 MVA for *Line 1–2* and 240 MVA for *Line 4–5*, respectively, are used. Furthermore, the capacity of each VSC is set to 200 MVA. Allowed voltage ranges lie between 0.9 and 1.1 pu on the AC side, and between 0.95 and 1.05 pu on the DC side.

6.2 *Multi-area SC-D-OPF*

To show applicability of the distributed approach to N-1 secure and dynamic OPF problems, a small example is presented in the following. An SC-D-OPF dispatch optimization is performed for the scenario set $\mathcal{C} = \{0, \textit{Line 1–2}\}$ and $\mathcal{T} = \{1, \ldots, 4\}$. Storages and VSCs are allowed curative control, that is, operating

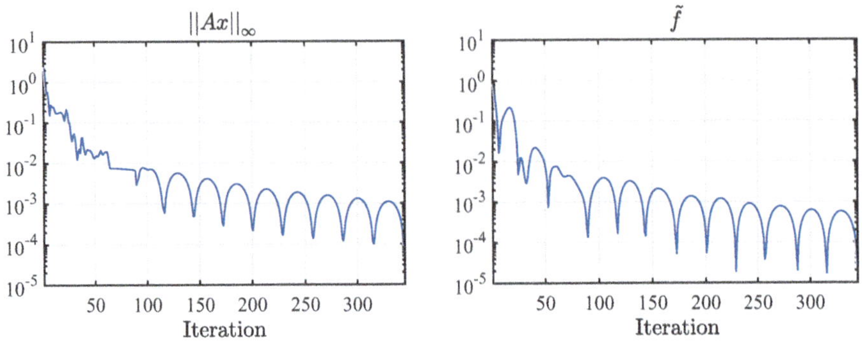

Fig. 4 Convergence behavior of distributed SC-D-OPF in the 5-bus system with 4 time steps and 1 contingency. Left: consensus error, right: cost error. Both relative to central solution

set points are allowed to move freely to a new set point after an outage. Contrarily, conventional generators are operated preventively, that is, the power set point must be valid both before and after the outage. However, to cope with changing system losses after an outage, a deviation of 1 % of installed capacity is allowed.

We use ADMM settings $\rho = 500$, $\tau = 1.02$, $\Theta = 0.99$, $W(S) = 1$ and $W(V) = 100$.

General convergence behavior is shown in Fig. 4. Consensus error and objective suboptimality show similar behavior compared to conventional OPF cases. Consensus error ($||Ax||_\infty$) which depicts cross-border feasibility, falls below the threshold of 10^{-4} after 344 iterations. The deviation from centrally computed costs f^* is denoted with $\tilde{f} = |1 - f/f^*|$. The objective value error falls below 0.01 % when convergence is achieved.

The progress of storage variables is shown in Fig. 5 for all time steps after the outage. Dashed lines denote results from the centralized solution. Storage systems provide positive (*ESS 2*) and negative (*ESS 1*) power reserve after the outage. The amount of reserve is increased with time for a growing network stress level, but energy levels remain within limits. The distributed solution is represented by solid lines. Those oscillate around and eventually approach the target levels.

7 Conclusion

This paper presents an approach to calculate SC-D-OPF in a distributed manner. That is, each control area computes a local N-1 secure and dynamic optimal dispatch which respects boundary constraints from neighboring areas for each time step and topology. After iteratively exchanging information and updating the local solution, a consensus is found which allows for a feasible cross-border power flow for all possible scenarios. Additionally, the result is close to the central optimal solution. Optimal curative storage or VSC set points can be determined for a contingency

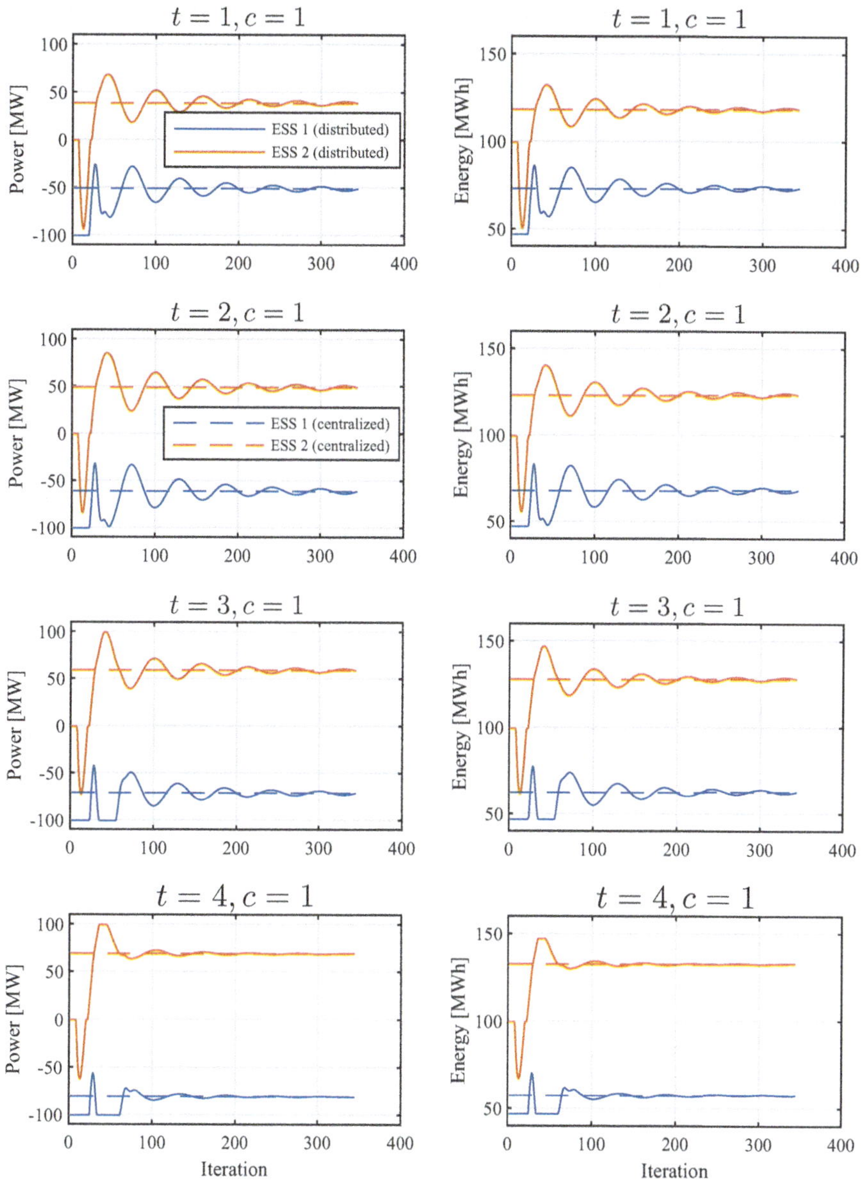

Fig. 5 Convergence of storage power and energy levels after the outage

scenario which is in a foreign area. In this work, we apply the method to a small test system – the next steps are to increase system size, time horizon and number of contingencies.

Appendix

Table 1 Line parameters with base power 100 MVA

Line	R [pu]	X [pu]	Total B [pu]
1–2	0.00281	0.0281	0.00712
1–4	0.00304	0.0304	0.00658
1–5	0.00064	0.0064	0.03126
2–3	0.00108	0.0108	0.01852
4–3	0.00297	0.0297	0.00674
5–4	0.00297	0.0297	0.00674
DC	0.0002	0	0

Table 2 Generator parameters. $(\underline{P_G}, \overline{P_G})$ in [MW] and $(\underline{Q_G}, \overline{Q_G})$ in [Mvar]

Generator	Node	$\underline{P_G}$	$\overline{P_G}$	$\underline{Q_G}$	$\overline{Q_G}$
G1	1	0	170	−127.5	127.5
G2	3	0	520	−390	390
G3	4	0	200	−150	150

Table 3 Generator cost coefficients

		Cost coefficients	
Generator	$a_G[\frac{1}{MW^2}\frac{€}{h}]$	$b_G[\frac{1}{MW}\frac{€}{h}]$	$a_q[\frac{1}{Mvar^2}\frac{€}{h}]$
G1	0.010	15	0.001
G2	0.011	30	0.001
G3	0.012	40	0.001
RES	0	5	0.001

Table 4 Parameters for each energy storage system

Capacity	200 MWh
Maximal power	100 MW
Minimal energy	20 MWh
Maximal energy	180 MWh
Efficiency	95%
Costs	
– discharge	10 €/MWh
– charge	0 €/MWh

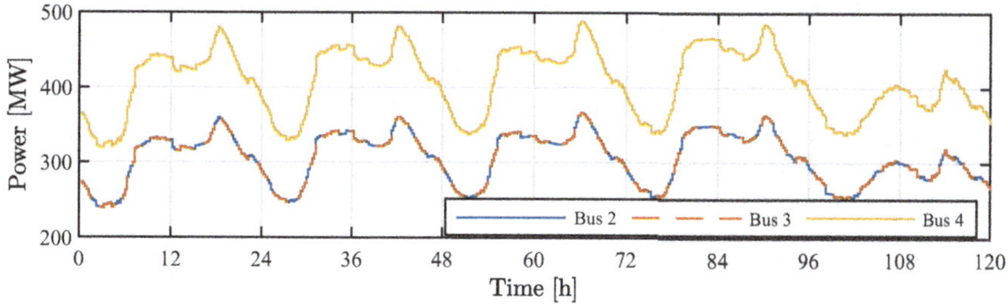

Fig. 6 Load profiles

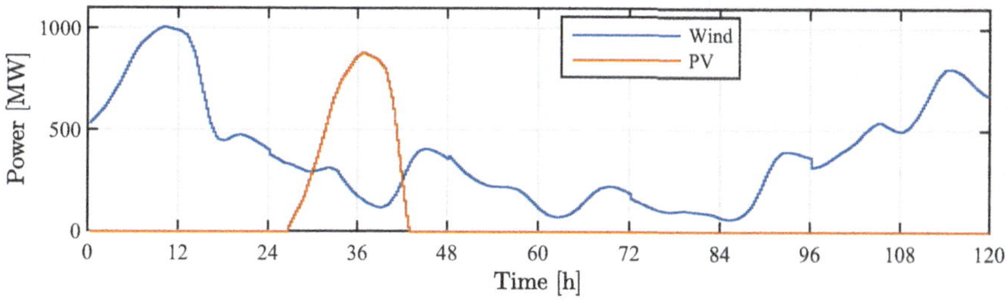

Fig. 7 RES profiles

Acknowledgements The authors kindly acknowledge the support for this work from the German Research Foundation (DFG) under the Project Number LE1432/14-2.

References

1. C.E. Murillo-Sanchez, R.D. Zimmerman, C.L. Anderson, R.J. Thomas, Decis. Support. Syst. **56**, 1 (2013). https://doi.org/10.1016/j.dss.2013.04.006
2. J. Eickmann, C. Bredtmann, A. Moser, in *Trends in Mathematics* (Springer International Publishing, 2017), pp. 47–63. https://doi.org/10.1007/978-3-319-51795-7_4
3. H. Sharifzadeh, N. Amjady, H. Zareipour, Electr. Power Syst. Res. **146**, 33 (2017). https://doi.org/10.1016/j.epsr.2017.01.011
4. A. Fuchs, J. Garrison, T. Demiray, in *2017 IEEE Manchester PowerTech*, 2017, pp. 1–6. https://doi.org/10.1109/PTC.2017.7981181
5. C.E. Murillo-Sanchez, R.D. Zimmerman, C.L. Anderson, R.J. Thomas, IEEE Trans. Smart Grid **4**(4), 2220 (2013). https://doi.org/10.1109/tsg.2013.2281001
6. A.J. Lamadrid, D.L. Shawhan, C.E. Murillo-Sánchez, R.D. Zimmerman, Y. Zhu, D.J. Tylavsky, A.G. Kindle, Z. Dar, IEEE Trans. Power Syst. **30**(2), 1064 (2015). https://doi.org/10.1109/TPWRS.2014.2388214
7. Y. Wen, C. Guo, H. Pandzic, D.S. Kirschen, IEEE Trans. Power Syst. **31**(1), 652 (2016). https://doi.org/10.1109/tpwrs.2015.2407054
8. N. Meyer-Huebner, M. Suriyah, T. Leibfried, in *2018 IEEE PES Innovative Smart Grid Technologies Conference Europe (ISGT-Europe)* (IEEE, 2018). https://doi.org/10.1109/isgteurope.2018.8571779

9. B.H. Kim, R. Baldick, IEEE Trans. Power Syst. **15**(2), 599 (2000). https://doi.org/10.1109/59.867147
10. D.K. Molzahn, F. Dörfler, H. Sandberg, S.H. Low, S. Chakrabarti, R. Baldick, J. Lavaei, IEEE Trans. Smart Grid **8**(6), 2941 (2017). https://doi.org/10.1109/TSG.2017.2720471
11. S. Boyd, N. Parikh, E. Chu, B. Peleato, J. Eckstein, Found. Trends Mach. Learn. **3**(1), 1 (2011). https://doi.org/10.1561/2200000016
12. T. Erseghe, IEEE Trans. Power Syst. **29**(5), 2370 (2014). https://doi.org/10.1109/TPWRS.2014.2306495
13. T. Erseghe, EURASIP J. Adv. Signal Process. **2015**(1), 45 (2015). https://doi.org/10.1186/s13634-015-0226-x
14. J. Guo, G. Hug, O.K. Tonguz, IEEE Trans. Power Syst. **32**(5), 3842 (2017). https://doi.org/10.1109/TPWRS.2016.2636811
15. N. Meyer-Huebner, M. Suriyah, T. Leibfried, IEEE Trans. Power Syst. **34**(4), 2937 (2019). https://doi.org/10.1109/TPWRS.2019.2892240
16. K. Baker, G. Hug, X. Li, in *2012 IEEE Power and Energy Society General Meeting* (IEEE, 2012). https://doi.org/10.1109/pesgm.2012.6344712
17. S. Mhanna, G. Verbic, A.C. Chapman, CoRR abs/1704.03647 (2017). http://arxiv.org/abs/1704.03647
18. A. Engelmann, T. Mühlpfordt, Y. Jiang, B. Houska, T. Faulwasser, in *IFAC-PapersOnLine* (20th IFAC World Congress, 2017), vol. 50, pp. 5536–5541. https://doi.org/10.1016/j.ifacol.2017.08.1095

A Domain Decomposition Approach to Solve Dynamic Optimal Power Flow Problems in Parallel

Nils Schween, Philipp Gerstner, Nico Meyer-Hübner, Viktor Slednev, Thomas Leibfried, Wolf Fichtner, Valentin Bertsch, and Vincent Heuveline

Abstract We propose a parallel solver for linear systems of equations arising from the application of Primal Dual Interior Point methods to Dynamic Optimal Power Flow problems. Our solver is based on the Generalised Minimal Residual method in combination with an additive Schwarz domain decomposition method as preconditioner. This preconditioner exploits the structure of Dynamic Optimal Power Flow problems which, after linearization, is given as a matrix with large diagonal blocks and only a few off-diagonal elements. These elements correspond to intertemporal couplings due to ramping and energy storage constraints and are partially neglected in order to solve the problem in parallel. We test our method on a large-scale optimisation problem and show that a parallel speedup can be obtained.

Keywords Dynamical OPF · Multi-period OPF · Time domain decomposition · Additive Schwarz method · Preconditioner

N. Schween · P. Gerstner · V. Heuveline (✉)
Heidelberg Institute for Theoretical Studies (HITS) and Engineering Mathematics and Computing Lab (EMCL), Heidelberg University, Heidelberg, Germany
e-mail: nils.schween@uni-heidelberg.de; philipp.gerstner@uni-heidelberg.de; vincent.heuveline@uni-heidelberg.de

N. Meyer-Hübner · T. Leibfried
Institute of Electric Energy Systems and High-Voltage Technology, Karlsruhe Institute of Technology (KIT), Karlsruhe, Germany

V. Slednev · W. Fichtner
Chair of Energy Economics, Institute for Industrial Production, Karlsruhe Institute of Technology (KIT), Karlsruhe, Germany

V. Bertsch
Department of Energy Systems Analysis, Institute of Engineering Thermodynamics, German Aerospace Center (DLR) and Institute for Building Energetics, Thermotechnology and Energy Storage, University of Stuttgart, Stuttgart, Germany

V. Bertsch et al. (eds.), *Advances in Energy System Optimization*,
Trends in Mathematics, https://doi.org/10.1007/978-3-030-32157-4_4

1 Introduction

While today's power grid infrastructure has been designed for centralised power generation in conventional power plants, the ever increasing share of renewable energy sources (*RES*) leads to an increasingly uncertain, volatile and decentralised generation. In order to ensure a reliable grid operation and to maintain today's security of supply, it is necessary to develop methods to simulate accurately a power grid at high temporal resolutions. This can be done by efficient methods for power grid optimisation, including an accurate consideration of the non-linear and non-convex AC power flow constraints. And these are to be improved.

The task to find the optimal operating state of a given power grid, also known as Optimal Power Flow (*OPF*), is formalized as the minimisation of a cost function with respect to a vector of continuous optimisation variables like the generated active power and the generated reactive power. Dynamic Optimal Power Flow (*DOPF*) considers several OPF problems, each corresponding to one specific point in time. In this case, the power grid's power demand is time-dependent and additional constraints must be taken into account. These constraints couple some of the optimisation variables corresponding to different time steps. Among others, these couplings are introduced by energy storage facilities and ramping constraints for conventional power plants [20]. Since DOPF is a large-scale non-linear and non-convex optimisation problem, solving it in parallel is of crucial importance.

For this kind of problems Primal Dual Interior Point Methods (*PDIPM*) have proven to be among the most powerful algorithms. The main computational effort, when PDIPM is applied, lies in solving linear systems of equations [14]. And this a task suitable to be carried out in parallel on multi-core CPUs.

There exist a few works on solving linear systems, that arise from PDIPM applied to DOPF, in parallel. One approach is based on parallel matrix factorization by means of Schur complement techniques [6, 20]. Other strategies use Benders Decomposition to decompose the complete optimisation problem into smaller ones [1].

Our contribution is the use of a *Schwarz Domain Decomposition Method* as preconditioner for Krylov-type iterative methods for solving these linear systems of equations in parallel. The original Schwarz method was formulated in 1870 as theoretical tool for proving existence of elliptic partial differential equations (*PDEs*) on complicated domains [17]. Later on, modifications of it have been used as stand-alone iterative methods for solving PDEs and have become a standard technique for preconditioning Krylov-type methods in the context of PDEs [18]. We apply this technique to DOPF by decomposing the time domain of interest into several smaller subdomains, which allows the use of multiple processors for computation.

2 Dynamic Optimal Power Flow

We will now formulate the DOPF problem in a mathematically rigorous way. This formulation has already been stated in more detail by several authors, e.g. in [12, 20] and [16]. We would like to mention that we consider a continuous formulation of the DOPF problem, i.e. one without any discrete decision variables. A non-linear mixed-integer formulation of an OPF is considered, for example, in [19].

2.1 *Formulation of the DOPF*

In order to formulate the DOPF problem for a given time period of interest $[0, T]$, we consider a uniform partition $0 = T_1 < T_2 < \ldots < T_{N_T} = T$ with constant step size $\tau = T_t - T_{t-1}$ for all $t \in \mathbf{T} \setminus \{1\}$, where $\mathbf{T} := \{1, \ldots, N_T\}$.

The power grid consisting of $N_\mathbf{B}$ nodes denoted by $\mathbf{B} := \{1, \ldots, N_\mathbf{B}\}$ and $N_\mathbf{E}$ transmission lines denoted by $\mathbf{E} \subset \mathbf{B} \times \mathbf{B}$, is described by an admittance matrix

$$\begin{aligned} &Y = G + jB \in \mathbb{C}^{N_\mathbf{B} \times N_\mathbf{B}} \\ &\text{with } G, B \in \mathbb{R}^{N_\mathbf{B} \times N_\mathbf{B}} \text{ and } j = \sqrt{-1}. \end{aligned} \tag{1}$$

The admittance matrix is symmetric, i.e. $Y = Y^T$, and furthermore $Y_{lk} = Y_{kl} \neq 0$ if and only if there is a branch connecting node k and l, i.e., $(k, l) \in \mathbf{E}$ and $(l, k) \in \mathbf{E}$. Therefore, Y is a sparse matrix for most real world power grids.

The complex voltage at node $k \in \mathbf{B}$ for time step $t \in \mathbf{T}$ is given as the complex number

$$V_k^t = E_k^t + jF_k^t \tag{2}$$

with real part $E_k^t \in \mathbb{R}$ and imaginary part $F_k^t \in \mathbb{R}$. We define $E^t := [E_k^t]_{k \in \mathbf{B}}$ and $F^t := [F_k^t]_{k \in \mathbf{B}}$.

Moreover, we define the following sets, which number:

$\mathbf{C} := \{1, \ldots, N_\mathbf{C}\}$	conventional power plants
$\mathbf{R} := \{1, \ldots, N_\mathbf{R}\}$	renewable energy sources
$\mathbf{S} := \{1, \ldots, N_\mathbf{S}\}$	storage facilities
$\mathbf{LS} := \{1, \ldots, N_\mathbf{LS}\}$	load shedding possibilities
$\mathbf{DC} := \{1, \ldots, N_\mathbf{DC}\}$	generators modelling DC lines
$\mathbf{V} := \{1, \ldots, N_\mathbf{V}\}$	generators modelling import/export

The resulting vector of variables corresponding to *active* power and energy for time step t is given by

$$P^t = \begin{pmatrix} [P^t_{g,i}]_{i\in\mathbf{C}} \\ [P^t_{r,i}]_{i\in\mathbf{R}} \\ [P^t_{sg,i}]_{i\in\mathbf{S}} \\ [P^t_{sl,i}]_{i\in\mathbf{S}} \\ [SOC^t_i]_{i\in\mathbf{S}} \\ [P^t_{l,i}]_{i\in\mathbf{LS}} \\ [P^t_{dc+,i}]_{i\in\mathbf{DC}} \\ [P^t_{dc-,i}]_{i\in\mathbf{DC}} \\ [P^t_{ex,i}]_{i\in\mathbf{V}} \\ [P^t_{im,i}]_{i\in\mathbf{V}} \end{pmatrix} \begin{matrix} \text{(injected conventional power)} \\ \text{(injected renewable power)} \\ \text{(injected storage power)} \\ \text{(extracted storage power)} \\ \text{(stored energy)} \\ \text{(load sheds)} \\ \text{(injected power by DC lines)} \\ \text{(extracted power by DC lines)} \\ \text{(power export)} \\ \text{(power import)} \end{matrix} \tag{3}$$

with $P^t \in \mathbb{R}^{N_P}$.

In a similar way, the vector of reactive power injections is defined by

$$Q^t = \begin{pmatrix} [Q^t_{g,i}]_{i\in\mathbf{C}} \\ [Q^t_{r,i}]_{i\in\mathbf{R}} \\ [Q^t_{sg,i}]_{i\in\mathbf{S}} \\ [Q^t_{sl,i}]_{i\in\mathbf{S}} \\ [Q^t_{l,i}]_{i\in\mathbf{LS}} \end{pmatrix} \in \mathbb{R}^{N_Q}. \tag{4}$$

Every variable related to active and reactive power can be assigned to a specific node of the power grid by means of the power injection and extraction matrices $C_P \in \{-1, 0, 1\}^{N_\mathbf{B}\times N_P}$ and $C_Q \in \{-1, 0, 1\}^{N_\mathbf{B}\times N_Q}$ [21, p.23]. Since the states of charge (SOC) of storage facilities do not directly influence the power flow equations, the corresponding rows in C_P are equal to the zero vector.

2.1.1 Constraints

At first, we consider the equality constraints. We begin with the power flow equations: Denoting the active and reactive power load for time step t by L^t_p and L^t_q, respectively, and assigning the loads to the nodes with the help of the matrix $C_L \in \{0, 1\}^{N_\mathbf{B}\times N_L}$, the AC power flow equations are given by [22]

$$C_P P^t - C_L L^t_p - Pf_p(E^t, F^t) = 0 \in \mathbb{R}^{N_\mathbf{B}}, \tag{5}$$

$$C_Q Q^t - C_L L^t_q - Pf_q(E^t, F^t) = 0 \in \mathbb{R}^{N_\mathbf{B}}, \tag{6}$$

where

$$Pf_{p,k}(E,F) = \sum_{l=1}^{N_\mathbf{B}} E_k(G_{kl}E_l - B_{kl}F_l) + F_k(G_{kl}F_l + B_{kl}E_l),$$

$$Pf_{q,k}(E,F) = \sum_{l=1}^{N_\mathbf{B}} F_k(G_{kl}E_l - B_{kl}F_l) - E_k(G_{kl}F_l + B_{kl}E_l),$$

which describe the sum of power flows at node k. Note that we dropped the index t to improve readability.

In the following, we abbreviate the equations (5) and (6) with the help of the following "symbolic" equation:

$$AC(E^t, F^t, P^t, Q^t) = 0. \tag{7}$$

Besides the power flow equations (7), one has to take into account an equality constraint for the slack bus, given by $F_1^t = 0$ for all t, and for generators $k, l \in \mathbf{DC}$ modelling a DC line $\overline{kl}$:

$$P^t_{dc+,k} = -\nu_{DC} P^t_{dc-,l} \text{ and } P^t_{dc+,l} = -\nu_{DC} P^t_{dc-,k}\,. \tag{8}$$

Here, line losses are taken into account by means of the factor $\nu_{DC} \in (0, 1)$.

Moreover, an additional set of equality constraints has to be imposed for modelling the state of charge, SOC_i^t, of storage facilities. These constraints are given by

$$SOC_i^t = SOC_i^{t-1} + \tau(\nu_l P^t_{sl,i} - \nu_g^{-1} P^t_{sg,i}), \quad \text{for all } i \in \mathbf{S} \tag{9}$$

with factors $\nu_l, \nu_g \in (0, 1)$ describing the efficiency of the loading process $P^t_{sl,i}$ and generation process $P^t_{sg,i}$, respectively.

Secondly, we take a look at the inequality constraints. Due to technical restrictions arising from power plants, renewable energy sources, storage facilities and DC transmission lines, it's necessary to impose lower and upper bounds for active and reactive power injections:

$$P^t_{min} \leq P^t \leq P^t_{max} \quad \text{and} \quad Q_{min} \leq Q^t \leq Q_{max}\,. \tag{10}$$

In our application the only bounds, which are time dependent, correspond to the power injected by renewable energy sources, i.e. P_r^t. Also the node voltages and the active power flow over transmission lines, denoted by p_{kl}, must be constrained:

$$U_{min}^2 \leq (E^t)^2 + (F^t)^2 \leq U_{max}^2 \tag{11}$$

$$p_{kl}(E_k^t, E_l^t, F_k^t, F_l^t) \leq Pf_{max,kl} \text{ for all } (k,l) \in \mathbf{E}\,. \tag{12}$$

Finally, we impose ramping constraints for conventional power plants to bound the rate of change for injected power between consecutive time steps by

$$|P_g^t - P_g^{t-1}| \leq \tau P_{max}^t \alpha \;\text{ for all } t \in \mathbf{T} \setminus \{1\}\,. \tag{13}$$

α is the ramping factor in percentage per unit time.

2.1.2 Cost Function

The cost function f accounts for expenditure and income related to active power processes for the complete time interval $[0, T]$ and is composed of contributions from each time step $t \in \mathbf{T}$ [5]:

$$f(P) = \sum_{t=1}^{N_T} \tau f_t(P^t) \tag{14}$$

with

$$\begin{aligned} f_t(P^t) = &\sum_{i \in \mathbf{C}} (a_{\mathbf{C},i,2}(P_{g,i}^t)^2 + a_{\mathbf{C},i,1} P_{g,i}^t) \\ &+ \sum_{i \in \mathbf{R}} (a_{\mathbf{R},i,2}(P_{r,i}^t)^2 + a_{\mathbf{R},i,1} P_{r,i}^t) \\ &+ \sum_{i \in \mathbf{V}} (a_{ex,i,2}(P_{ex,i}^t)^2 + a_{ex,i,1} P_{ex,i}^t) \\ &+ \sum_{i \in \mathbf{V}} (a_{im,i,2}(P_{im,i}^t)^2 + a_{im,i,1} P_{im,i}^t) \\ &+ \sum_{i \in \mathbf{LS}} (a_{\mathbf{LS},i,2}(P_{l,i}^t)^2 + a_{\mathbf{LS},i,1} P_{l,i}^t) \end{aligned} \tag{15}$$

and coefficients $a_{*,*,*} \in \mathbb{R}$.

2.1.3 Optimisation Problem

With the definitions of the previous sections at hand, the DOPF problem can be stated in an abstract way:

$$
\begin{aligned}
&\min_{x} \ f(x) \text{ s.t.} \\
&\qquad g_I^t(x^t) = 0, \qquad && \text{for all } t \in \mathbf{T} \\
&\qquad g_S^t(x^t, x^{t-1}) = 0, && \text{for all } t \in \mathbf{T} \setminus \{1\} \\
&\qquad h_I^t(x^t) \le 0, && \text{for all } t \in \mathbf{T} \\
&\qquad h_R^t(x^t, x^{t-1}) \le 0, && \text{for all } t \in \mathbf{T} \setminus \{1\}
\end{aligned}
\tag{16}
$$

where the vector of optimisation variables is given by

$$
x = [x^t]_{t \in \mathbf{T}} \in \mathbb{R}^{n^x} \text{ and } x^t = \begin{pmatrix} E^t & F^t & P^t & Q^t \end{pmatrix} \in \mathbb{R}^{n^{x,t}}. \tag{17}
$$

Moreover, we used and will use the following definitions: $n^{x,t} := 2N_{\mathbf{B}} + N_P + N_Q$ and $n^x := N_T \, n^{x,t}$ and, as of now, let $n^{\lambda,t}$ be defined as the number of equality constraints corresponding to time step t, i.e. $\lambda^t \in \mathbb{R}^{n^{\lambda,t}}$ with $\lambda^t = (\lambda_I^t, \lambda_S^t)$ denoting the Lagrangian multipliers corresponding to g_I^t and g_S^t, respectively. Analogously, $\mu^t = (\mu_I^t, \mu_R^t) \in \mathbb{R}^{n^{\mu,t}}$ denotes the Lagrangian multipliers for the inequality constraints h_I^t and h_R^t. Consequently, we define $n^\lambda := N_T n^{\lambda,t}$ and $n^\mu := N_T n^{\mu,t}$.

Note that the ramping constraints h_R given by (13) and the storage constraints g_S given by (9) establish the only couplings between the variables corresponding to different time steps. Without them, a solution to (16) could be obtained by solving N_T independent OPF problems. The optimisation problem (16) is non-linear due to the AC equations and due to the voltage and line flow inequality constraints.[1] Moreover, the latter ones and the AC equations are the sources of non-convexity in (16).

To simplify the notation in the following sections, we abbreviate the optimisation problem (16) to

$$
\min_{x} f(x) \quad \text{s.t.} \ \ g(x) = 0, \ h(x) \le 0 \tag{18}
$$

with quadratic functions

$$
f\colon \mathbb{R}^{n^x} \to \mathbb{R}, \ g\colon \mathbb{R}^{n^x} \to \mathbb{R}^{n^\lambda}, \ h\colon \mathbb{R}^{n^x} \to \mathbb{R}^{n^\mu}. \tag{19}
$$

[1] We have not explicitly stated an equation for the power flow p_{kl} through the transmission line (k, l). Please refer to [13].

3 Primal Dual Interior Point Method for DOPF

In this section, we briefly describe the application of the Primal Dual Interior Point Method (*PDIPM*) to a DOPF problem. The description is in line with the implementation found in the MATLAB package MATPOWER [21].

3.1 The PDIPM Algorithm

For the general optimisation problem (18), the corresponding Lagrangian function is

$$
\begin{aligned}
L\colon \mathbb{R}^{n^x} \times \mathbb{R}^{n^\lambda} \times \mathbb{R}^{n^\mu} &\to \mathbb{R} \\
(x, \lambda, \mu) &\mapsto f(x) + \lambda^T g(x) + \mu^T h(x).
\end{aligned} \tag{20}
$$

Assuming additional constraint qualifications, e.g., the linear independence constraint qualification (*LICQ*) [14], for every local minimum x^* of (18), there exist corresponding Lagrangian multipliers λ^*, μ^* such that (x^*, λ^*, μ^*) are in accord with the Karush-Kuhn-Tucker (*KKT*) conditions [5]:

$$
\mathbf{F}_0(x, s, \lambda, \mu) := \begin{pmatrix} \nabla_x L(x, \lambda, \mu) \\ g(x) \\ h(x) + s \\ S\mu \end{pmatrix} = 0, \quad s, \mu \geq 0, \tag{21}
$$

where $s \in \mathbb{R}^{n^\mu}$ is a vector containing slack variables and $S \in \mathbb{R}^{n^\mu \times n^\mu}$ denotes a diagonal matrix whose diagonal elements are $S_{kk} = s_k$.

The PDIPM is an algorithm to find solutions to Eq. (21). Mathematically it solves a slightly modified version of this equation by applying to it Newton's method. The modification consists in adding a "term", which decreases with iteration index k. This term keeps the slack variables s away from zero and, thus, the inequality constraints inactive. This means that solutions $(x^{(k)}, s^{(k)}, \lambda^{(k)}, \mu^{(k)})$ to this modified version of Eq. (21) are inside the feasible region and *not* on its surface, hence, the method is called interior point method. More rigorously formulated, we find

$$
\begin{aligned}
\mathbf{F}_\gamma(x, s, \lambda, \mu) := \mathbf{F}_0(x, s, \lambda, \mu) - \begin{pmatrix} 0 \; 0 \; 0 \; \gamma e \end{pmatrix}^T &= 0, \\
s, \mu &\geq 0,
\end{aligned} \tag{22}
$$

for a sequence of *barrier parameters*

$$\gamma_k := \sigma \frac{(\mu^{(k-1)})^T s^{(k-1)}}{n^\mu}$$

where $\sigma = 0.1$ is the *centering parameter* and $(\mu^{(k-1)})^T s^{(k-1)}$ is called *complementary gap*. Additionally, $e = (1, \ldots, 1) \in \mathbb{R}^{n^\mu}$. For details we refer the inclined reader to [5, 8] and [21].

Remark A very basic PDIPM is applied, i.e. there is no globalisation strategy. Since we focus on parallel linear algebra techniques for DOPF problems, a locally convergent algorithm is sufficient to our purpose.

3.2 KKT Matrix in PDIPM

The application of Newton's method to Eq. (22) yields iterations in which the following linear systems of equations must be solved:

$$\nabla \mathbf{F}_{\gamma_k}(x^{(k)}, s^{(k)}, \lambda^{(k)}, \mu^{(k)})\Delta^{(k)} = -\mathbf{F}_{\gamma_k}(x^{(k)}, s^{(k)}, \lambda^{(k)}, \mu^{(k)}) \,. \tag{23}$$

Henceforth we omit the iteration index k. Assuming that $s, \mu > 0$, the Newton system (23) can be transformed into a reduced, symmetric saddle point form

$$\underbrace{\begin{pmatrix} \nabla^2_{xx} L + (\nabla h)^T \Sigma (\nabla h) & (\nabla g)^T \\ \nabla g & 0 \end{pmatrix}}_{=:\, A(x,s,\lambda,\mu)} \underbrace{\begin{pmatrix} \Delta x \\ \Delta \lambda \end{pmatrix}}_{=:\Delta_R} = - \underbrace{\begin{pmatrix} r_x \\ r_\lambda \end{pmatrix}}_{=:b} \tag{24}$$

with diagonal matrix $\Sigma = \operatorname{diag}\left(\frac{\mu_1}{s_1}, \ldots, \frac{\mu_{n^\mu}}{s_{n^\mu}}\right)$ and

$$r_x = \nabla_x L + (\nabla h(x))^T Z^{-1} \left(\gamma e + \operatorname{diag}(\mu) h(x)\right) ,$$
$$r_\lambda = g(x) \,.$$

The KKT matrices A, arising in every PDIPM iteration k, may become more and more ill-conditioned when a KKT point is approached. This is caused by the sequence of scaling matrices $\Sigma^{(k)}$. If strict complementary holds at the KKT point x^*, if $x^{(k)} \to x^*$ and if $h_i(x^*)$ is an active inequality, one can show that $\Sigma^{(k)}_{ii} \to \infty$. On the other hand, $\Sigma^{(k)}_{ii} \to 0$, if $h_i(x^*)$ is inactive [14].[2] For this reason, many IPM software packages (like IPOPT [7]) use direct methods such as

2

LDL^T-factorizations to solve the appearing linear systems. These methods are less sensitive to ill-conditioned matrices.

In contrast, iterative linear solvers like GMRES [15] are very sensitive to ill-conditioned matrices. A good preconditioner is needed to lower the matrices' condition numbers. In exchange, they offer a higher potential of parallelization and allow the use of inexact Interior Point methods, see Sect. 3.3.

The distribution of the spectrum of a matrix K, defined as

$$\sigma(K) := \{\lambda \in \mathbb{C},\ \lambda \text{ is an eigenvalue of } K\},$$

is an indicator of the convergence behaviour of GMRES applied to $Kv = b$. By rule of thumb, a spectrum which is clustered away from 0 is beneficial to the rate of convergence of GMRES. [11, Th. 4.62]

For general non-linear optimisation problems with the corresponding KKT matrix

$$K = \begin{pmatrix} H & B^T \\ B & 0 \end{pmatrix}, \quad H \in \mathbb{R}^{n^x \times n^x},\quad B \in \mathbb{R}^{n^\lambda \times n^x},\quad n^\lambda \leq n^x, \tag{25}$$

the so-called *constraint preconditioner* is an example of a preconditioner. It is given by

$$\tilde{K} = \begin{pmatrix} \tilde{H} & B^T \\ B & 0 \end{pmatrix}, \tag{26}$$

with $\tilde{H}$ being an approximation to H, that is easy to factorize, e.g., $\tilde{H} = \mathrm{diag}(H)$ [14].

One can show that $\sigma(\tilde{K}^{-1}K) = \{1\} \cup \tilde{\sigma}$ with $|\tilde{\sigma}| = n^x - n^\lambda$. Therefore, at most $n^x - n^\lambda$ eigenvalues of $\tilde{K}^{-1}K$ are not equal to 1. The distribution of these remaining eigenvalues depends on how well $\tilde{H}$ approximates H on the nullspace of B [10].[3]

In Sect. 4, we describe how to take advantage of the special structure given by (16) to construct a parallel preconditioner. Furthermore, we will see that there is a close relation between our proposed preconditioner and the constraint preconditioner defined above, see Sect. 4.4.

Proof (sketch) The KKT conditions require that $\mu_i^*(h_i(x^*) + s_i^*) = 0$. If $h_i(x^*)$ is active, then $h_i(x^*) = 0$ and $s_i^* = 0$. Upon strict complementarity follows that $\mu_i^* \neq 0$. Finally, if $x^{(k)} \to x^*$, then $\mu_i^{(k)} \to \mu_i^*$, $s_i^{(k)} \to s_i^*$ and $\Sigma_{ii}^{(k)} = \frac{\mu_i^{(k)}}{s_i^{(k)}} \to \infty$. □

[3] Let Z denote the nullspace of B, i.e. $BZ = 0$. Then $\tilde{H}$ approximates B on its nullspace, if $(Z^T\tilde{H}Z)^{-1}(Z^T HZ) \approx 1$.

3.3 Inexact PDIPM

In contrast to direct methods, iterative solvers allow solving linear systems with prescribed accuracy. One can exploit this fact by means of inexact Interior Point methods. Within these methods, the linear systems corresponding to the first PDIPM iterations are solved with a rather low accuracy and therefore a low number of iterations. As the PDIPM iterate approaches the exact solution, the accuracy of the linear solver is increased. We use the approach from [3] for determining the accuracy in each PDIPM iteration. Here, (23) has to be solved with residual $\tilde{r}^{(k)}$, i.e,

$$\begin{aligned}\nabla \mathbf{F}_{\gamma_k}(x^{(k)}, s^{(k)}, \lambda^{(k)}, \mu^{(k)})\Delta^{(k)} = &- \mathbf{F}_{\gamma_k}(x^{(k)}, s^{(k)}, \lambda^{(k)}, \mu^{(k)}) \\ &+ \tilde{r}^{(k)},\end{aligned}$$

that satisfies $\|\tilde{r}^{(k)}\|_2 \le \tilde{\eta}_k \frac{(\mu^{(k)})^T s^{(k)}}{n^\mu}$ with forcing sequence $\tilde{\eta}_k \in (0, 1)$. With this choice, the update step $\Delta^{(k)}$ satisfies

$$\begin{aligned}\nabla \mathbf{F}_0(x^{(k)}, s^{(k)}, \lambda^{(k)}, \mu^{(k)})\Delta^{(k)} = &- \mathbf{F}_0(x^{(k)}, s^{(k)}, \lambda^{(k)}, \mu^{(k)}) \\ &+ r^{(k)}\end{aligned}$$

with $\|r^{(k)}\|_2 \le (\gamma_k + \tilde{\eta}_k)\|\mathbf{F}_0(x^{(k)}, s^{(k)}, \lambda^{(k)}, \mu^{(k)})\|_2$. Therefore, "$\gamma_k + \tilde{\eta}_k$ can be viewed as forcing sequence of inexact Newton methods" [3]. In our numerical experiments, we set $\tilde{\eta}_k = 0.1$ for all k and computed the relative tolerance for the iterative solver as follows

$$\mathrm{rtol}^{(k)} = 0.1 \frac{{\mu^{(k)}}^T s^{(k)}}{n^\mu \|\mathbf{F}_{\gamma_k}(x^{(k)}, s^{(k)}, \lambda^{(k)}, \mu^{(k)})\|}.$$

We stopped to decrease the relative tolerance when it reached a value smaller than 10^{-10}.

3.4 Termination Criteria for PDIPM

We use the following criteria for monitoring the progress of PDIPM [21]:

$$c_{feas}(x, s, \lambda, \mu) := \frac{\max\{\|g(x)\|_\infty, \max_i\{h_i(x)\}\}}{1 + \max\{\|x\|_\infty, \|s\|_\infty\}} \tag{27}$$

$$c_{grad}(x, s, \lambda, \mu) := \frac{\|\nabla_x L(x, \lambda, \mu)\|_\infty}{1 + \max\{\|\lambda\|_\infty, \|\mu\|_\infty\}} \tag{28}$$

$$c_{comp}(x, s, \lambda, \mu) := \frac{(\mu^T s)}{1 + \|x\|_\infty} \tag{29}$$

$$c_{cost}(x, s, \lambda, \mu) := \frac{|f(x) - f(x^-)|}{1 + |f(x^-)|} \tag{30}$$

with x^- denoting the previous iterate.

4 Schwarz Domain Decomposition Method for DOPF

The key idea behind Schwarz domain decomposition methods applied to DOPF problems is the decomposition of the set of time steps $\mathbf{T}$ into several smaller subsets. Due to this decomposition, it is possible to define corresponding submatrices of the global KKT matrix defined in (24) that are smaller in size and therefore faster to factorize. Furthermore, the factorization of these submatrices can be done in parallel. After computing subsolutions corresponding to these submatrices, one can reconstruct an approximate solution to the global system (24), which is used as preconditioner within a Krylov-type iterative solver.

In Sect. 4.1 we introduce the mathematics, which describe the additive Schwarz Method (*ASM*) in the context of DOPF problems. In Sect. 4.2 we briefly describe some issues related to its implementation. Section 4.3 is intended to provide a deeper insight into the subproblems that have to be solved when the ASM is applied. Finally, we describe the relation between the ASM and the above mentioned constraint preconditioner in Sect. 4.4.

4.1 Mathematical Formulation of the Additive Schwarz Method

For the application of the ASM as a preconditioner for the KKT matrix $A(x, s, \lambda, \mu) \in \mathbb{R}^{(n^x+n^\lambda)\times(n^x+n^\lambda)}$ defined by (24), it is needed to decompose the set of time steps $\mathbf{T} = \{1, \ldots, N_T\}$ into q non-overlapping subdomains:

$$\mathbf{T} = \bigcup_{l=1}^{q} \tilde{\mathbf{T}}_l,$$
$$\text{with } \tilde{\mathbf{T}}_l \cap \tilde{\mathbf{T}}_k = \emptyset \text{ for } k \neq l \text{ and } \tilde{\mathbf{T}}_l := \{\tilde{t}_l^-, \tilde{t}_l^- + 1, \ldots, \tilde{t}_l^+\}.$$

Afterward, each subdomain $\tilde{\mathbf{T}}_l$ is expanded by additional s_{ol} time steps on both ends, yielding an overlapping decomposition of $\mathbf{T}$ (see Fig. 1):

$$\mathbf{T} = \bigcup_{l=1}^{q} \mathbf{T}_l, \; \mathbf{T}_l := \{t_l^-, t_l^- + 1, \ldots, t_l^+\},$$

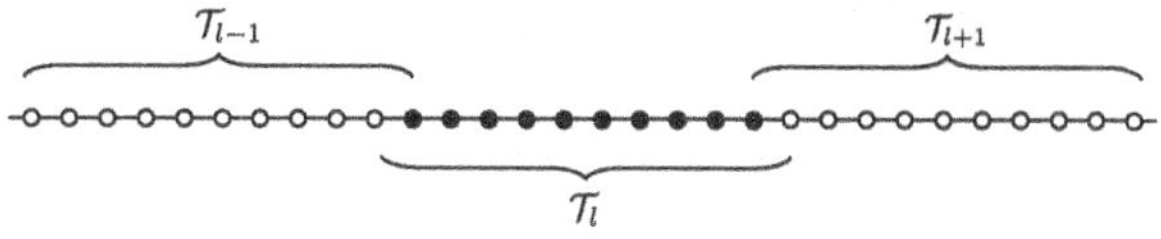

Fig. 1 Decomposition of time steps with overlap $s_{ol} = 1$

with

$$t_l^- = \begin{cases} \tilde{t}_l^- - s_{ol}, & l > 1 \\ \tilde{t}_l^-, & l = 1 \end{cases}, \quad t_l^+ = \begin{cases} \tilde{t}_l^+ + s_{ol}, & l < q \\ \tilde{t}_l^+, & l = q \end{cases}$$

Typically, $s_{ol} \in \{1, 2, 3\}$.

Furthermore, we use the definitions of Sect. 2.1.3 to introduce

$$n_l^x := \sum_{t \in \mathbf{T}_l} n^{x,t},$$

$$n_l^\lambda := \sum_{t \in \mathbf{T}_l} n^{\lambda,t}$$

$$n_l := n_l^x + n_l^\lambda,$$

$$n := n^x + n^\lambda .$$

Henceforward, all expressions involving ˜ refer to the non-overlapping subdomains $\tilde{\mathbf{T}}_l$.

When restricting the optimisation variables to the components, which belong to $\mathbf{T}_l$, we write

$$x_{[l]} = [x^t]_{t \in \mathbf{T}_l} \text{ and} \tag{31}$$

$$\lambda_{[l]} = [\lambda^t]_{t \in \mathbf{T}_l} = \left[\begin{pmatrix} \lambda_I^t \\ \lambda_S^t \end{pmatrix} \right]_{t \in \mathbf{T}_l} . \tag{32}$$

The combination of $x_{[l]}$ and $\lambda_{[l]}$ yields the following definition:

$$\Delta_{R[l]} := \begin{pmatrix} \Delta x \\ \Delta \lambda \end{pmatrix}_{[l]} := \begin{pmatrix} \Delta x_{[l]} \\ \Delta \lambda_{[l]} \end{pmatrix} \in \mathbb{R}^{n_l} . \tag{33}$$

This vector contains the components of the solution vector of (24) belonging to the time steps in $\mathbf{T}_l$. Moreover, let $R_l \in \{0, 1\}^{n_l \times n}$ be a *restriction matrix* corresponding to the subset $\mathbf{T}_l$ defined by its action:

$$R_l \begin{pmatrix} \Delta x \\ \Delta\lambda \end{pmatrix} = \begin{pmatrix} \Delta x \\ \Delta\lambda \end{pmatrix}_{[l]}, \quad \text{where } R_l = \begin{pmatrix} R_l^x & 0 \\ 0 & R_l^\lambda \end{pmatrix}. \tag{34}$$

The matrices R_l^x and R_l^λ are composed of either zero or identity blocks. As an example, consider the case of $l \neq 1$ and $l \neq q$:

$$\begin{aligned} R_l^x &= \begin{pmatrix} 0 & I_{n_l^x} & 0 \end{pmatrix} \in \{0, 1\}^{n_l^x \times n^x}, \\ R_l^\lambda &= \begin{pmatrix} 0 & I_{n_l^\lambda} & 0 \end{pmatrix} \in \{0, 1\}^{n_l^\lambda \times n^\lambda}. \end{aligned} \tag{35}$$

Note that the size of the zero matrices varies with each restriction matrix and each l. Analogously, we define $\tilde{R}_l$ as restriction operator corresponding to the non-overlapping subdomain $\tilde{\mathbf{T}}_l$.

With these definitions at hand, it is possible to extract submatrices A_l from the KKT matrix A by multiplying it with R_l:

$$A_l := R_l A R_l^T. \tag{36}$$

In Sect. 4.3, we explore the structure of these submatrices.

On the important assumption, that all submatrices A_l are *non-singular*, we are able to formulate the multiplicative Schwarz Method (*MSM*) for approximately solving $A\Delta_R = b$. We do this in the form of an algorithm [18]:

Algorithm 1 Multiplicative Schwarz method

Set $\Delta_R^0 = 0$
for $l = 1, \ldots, q$ **do**
$\Delta_R^l = \Delta_R^{l-1} + R_l^T A_l^{-1} R_l \left(b - A\Delta_R^{l-1}\right) = \Delta_R^{l-1} + \delta_l$
end for
Return Δ_R^q

In every iteration l, the current error with respect to the exact solution Δ_R, given by

$$e^{l-1} = \Delta_R - \Delta_R^{l-1},$$

satisfies

$$Ae^{l-1} = r^{l-1} \text{ with residual vector } r^{l-1} = b - A\Delta_R^{l-1}.$$

By restricting r^{l-1} to subdomain $\mathbf{T}_l$ and solving with A_l^{-1}, one obtains an approximation

$$\hat{e}^l := A_l^{-1} R_l r^{l-1}$$

to the exact error $R_l e^{l-1}$ on subdomain $\mathbf{T}_l$. Prolongation with R_l^T yields a correction $\delta_l = R_l^T \hat{e}^l$ to the current approximation Δ_R^{l-1}. Thus, the multiplicative Schwarz method can be seen as "defect correction algorithm".[4]

Unfortunately, the the MSM is a sequential algorithm. In order to parallelize it, one omits residual updates in each iteration, yielding the ASM [18]:

Algorithm 2 Additive Schwarz method $\Delta_R^q = \text{ASM}(b)$

Set $\Delta_R^0 = 0$
for $l = 1, \ldots, q$ **do**
  $\Delta_R^l = \Delta_R^{l-1} + R_l^T A_l^{-1} R_l b$
end for
Return Δ_R^q

which can be written as

$$\Delta_R^q = M_{ASM} b \quad \text{with } M_{ASM} := \sum_{l=1}^{q} R_l^T A_l^{-1} R_l \,.$$

The right preconditioned linear system for solving $A\Delta_R = b$ is then given by

$$A M_{ASM} u = b, \; \Delta_R = M_{ASM} u \,. \tag{37}$$

Since M_{ASM} is symmetric but in general not positive definite, we use the Generalized Minimal Residual (*GMRES*) method for solving the non-symmetric system (37). In general, the ASM computes a less accurate approximation to the solution of $A\Delta_R = b$ and therefore needs an increased number of GMRES iterations compared to the MSM given by Algorithm 1. However, the ASM offers a higher potential of parallelization, which results usually in better parallel scaling.

[4]In contrast to "classical defect correction" algorithms, there is no converging sequence $\left(\Delta_R^l\right)_{l\in\mathbb{N}} \to \Delta^R$. Nonetheless, the algorithm's construction implies that $\Delta_R^q \approx \Delta^R$.

4.2 Implementation

In our implementation, we use one processor per subdomain $\mathbf{T}_l$ and distribute A such that every processor stores $\tilde{R}_l A$ in its local memory. $\tilde{R}_l A$ contains the non-overlapping portion of rows of $A_l = R_l A R_l^T$.

To set up M_{ASM} once, every processor first has to form its local submatrix A_l, i.e., in this step the overlapping part of A_l has to be communicated to processor l by processor $l-1$ and $l+1$. Afterward, each processor computes an LU-factorization of A_l, i.e., $A_l = L_l U_l$. This step doesn't involve any inter-processor communication and can be done in parallel.

The employment of the ASM as a preconditioner inside the GMRES method, requires to compute $M_{ASM} v^{(k)}$ in each iteration k. $v^{(k)}$ denotes the k-th basis vector of the Krylov subspace $\mathscr{K}_k(AM_{ASM}, b)$. On the assumption that the basis vector's data layout is the same as the matrix's one, i.e. every process stores $v_l^{(k)} := \tilde{R}_l v^{(k)}$, communication with process $l-1$ and $l+1$ is required to multiply M_{ASM} with $v^{(k)}$.[5] The computation of $A_l^{-1} v_l^{(k)}$ is done by one forward substitution with L_l and a backward substitution with U, respectively. This step does not involve any communication. After this, the local solution $A_l^{-1} v_l^{(k)}$ is prolonged back to obtain the global result of the matrix vector product $M_{ASM} v^{(k)}$. This again requires communicating the overlapping part of $v_l^{(k)}$ to process $l-1$ and $l+1$. Finally, it's necessary to multiply A with $M_{ASM} v^{(k)}$. Due to A's data layout and its off-diagonal elements, corresponding to intertemporal couplings, additional communication is required.

One can further improve the performance of ASM by using the so-called *restricted* version of ASM [4], which is given by

$$M_{rASM} := \sum_{l=1}^{p} \tilde{R}_l^T A_l^{-1} R_l \,. \tag{38}$$

For this preconditioner, just the non-overlapping part of the local solution $v_l^{(k)}$ is prolonged instead of the entire (overlapping) vector. Experiments show a beneficial behaviour in terms of GMRES iterations compared to standard ASM [4]. Furthermore, prolongation by $\tilde{R}_l$ doesn't involve any communication.

To further stabilise the LU-factorization, a small regularisation parameter $\epsilon = 10^{-8}$ is added to the KKT matrix A, i.e. $A + \epsilon I$.

[5]The reader may wonder why we have to make an assumption about the basis vectors' data layout: Our GMRES implementation is part of the external library PETSc and a quick glance at their source code did not reveal to us how they manage the according data.

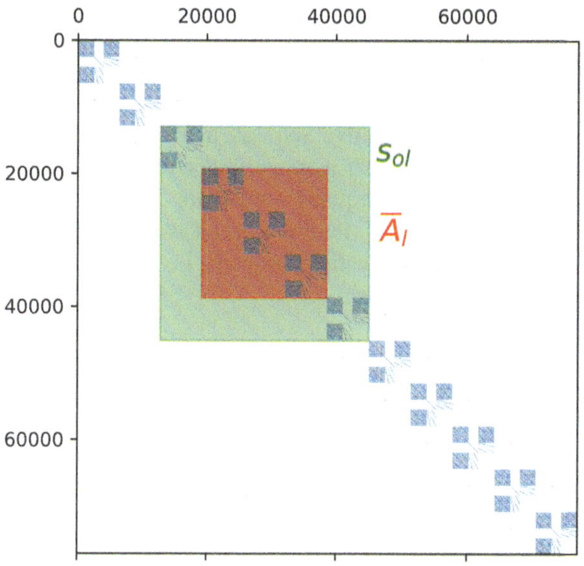

Fig. 2 Permuted KKT matrix A with $N_T = 12, s_{ol} = 1, q = 4$ and $\bar{A}_l = P_l A_l P_l{}^T$

4.3 Structure of Local KKT Matrices

In this section, we investigate the structure of the local submatrices A_l defined in Eq. (36). A good way to get an idea of their structure is to permute the KKT matrix A in such a way that all rows and columns belonging to *one* time step are grouped together. This yields a block diagonal matrix with some off-diagonal elements. Every block represents one time step and the off-diagonal elements arise due to the time-dependent constraints. Extracting the time steps belonging to the subdomain $\mathbf{T}_l$ results in a permuted version of the submatrix A_l.[6] Its structure is exactly the same as the one of the original KKT matrix A. If the off-diagonal elements are taken into account as "boundary conditions", it is possible to interpret the submatrix A_l as representing an optimisation problem in its own, i.e. the original dynamical OPF problem is reduced to the subdomain $\mathbf{T}_l$. Figure 2 tries to demonstrate this. As a consequence the submatrices A_l tend to be ill-conditioned.

[6]Note that this extraction cannot be done by multiplying A with R_l and $R_l{}^T$.

4.4 Relationship Between the ASM and the Constraint Preconditioner

To see a relation between the additive Schwarz method and the constraint preconditioner, as introduced in Sect. 3.2, it is necessary to consider a dynamical OPF problem without time-dependent equality constraints. In our case, we had to drop the storage facilities g_S^t to do this. Furthermore, it is important to consider a non-overlapping decomposition of the time domain, i.e. $s_{ol} = 0$. In this case, the corresponding ASM preconditioner,

$$\tilde{M}_{ASM} = \sum_{l=1}^{q} \tilde{R}_l^T \tilde{A}_l^{-1} \tilde{R}_l, \quad (\text{where } \tilde{A}_l := \tilde{R}_l A \tilde{R}_l^T \in \mathbb{R}^{\tilde{n}_l \times \tilde{n}_l}) \tag{39}$$

reduces to a block Jacobi method with omitted couplings between variables belonging to $\tilde{t}_l^+$ and $\tilde{t}_{l+1}^-$ for each subdomain l. Since there are only time-dependent inequality constraints left, the coupling between the variables belonging to different time steps has to arise in the $(1, 1)$-block of the KKT matrix, namely in the $(\nabla h)^T \Sigma (\nabla h)$ part. This means that multiplying $\tilde{M}_{ASM}$ with b is like solving a modified KKT system, or equivalently,

$$\tilde{M}_{ASM} = \begin{pmatrix} \tilde{H} & (\nabla g)^T \\ \nabla g & 0 \end{pmatrix}^{-1} . \tag{40}$$

$\tilde{H}$ denotes the KKT matrix's modified $(1, 1)$-block. Thus, $\tilde{M}_{ASM}$ has the form of a constraint preconditioner.

At this point, it also becomes clear that the ASM can only be interpreted as a constraint preconditioner, if there are no time-dependent equality constraints. If there existed time-dependent equality constraints, the $(2, 1)$- and $(1, 2)$-block of $\tilde{M}_{ASM}^{-1}$ would have changed too.

As pointed out in Sect. 3.2, 1 is an eigenvalue of $\tilde{M}_{ASM}A$ of multiplicity $2n^\lambda$ and the remaining $n^x - n^\lambda$ eigenvalues are solutions of a generalized eigenvalue problem of the following form [10]:

$$Z^T \underbrace{(\nabla_{xx}^2 \mathbf{L} + (\nabla h)^T \Sigma (\nabla h))}_{=:H} Zv = \lambda Z^T \tilde{H} Zv . \tag{41}$$

For a low number of subdomains q, one can expect $\tilde{H}$ to be a good approximation to H, leading to a clustered spectrum $\sigma(\tilde{M}_{ASM}A)$ close to one. However, the more subdomains, the more neglected couplings and the less accurate does $\tilde{H}$ approximate H. This results in a more scattered eigenvalue distribution of $\tilde{M}_{ASM}A$, but still a large number of eigenvalues are equal to 1.

5 Numerical Experiments

In this section, we present some results for our proposed method applied to a DOPF problem. In its first part, we discuss the parallel speedup of our solver. In the second one, we investigate the way in which the KKT matrix's spectrum $\sigma(A)$ changes with the iterations of the interior point method. The test grid, which we used for our experiments, represents a possible variant of the German Transmission Grid in the year 2023 and consists of 1215 nodes, 2247 AC lines, 8 DC lines, 181 conventional power plants, 355 renewable energy sources and 67 storage facilities. For more details, we refer the reader to [12].

For solving (16), we use the PDIPM algorithm mips which is written in MATLAB code and part of MATPOWER [21]. In this algorithm, we replace the standard MATLAB backslash operator \ for solving linear systems by our own linear solver. Our solver applies a right preconditioned GMRES method. We use the ASM as the preconditioner. For computing the LU-factorizations of the local systems A_l, we use the software package Intel® MKL PARDISO. Our solver is written in C++ and makes use of the KSPGMRES and PCASM methods provided by PETSc [2], which is compiled in Release mode with the Intel compiler 17.0. The computation of the eigenvalues of the KKT matrices and the preconditioned KKT matrices have been done with SLEPc [9] and MATLAB. Inter-process communication is realised by means of the Message Passing Interface (MPI). The numerical experiments were carried out on the BwForCluster MLS&WISO Production at Heidelberg University with two Intel Xeon E5-2630v3 processors i.e. 16 CPU cores at 2.4 GHz per node, and 64 GB working memory.

5.1 Parallel Speedup

In our numerical experiments, we concatenated the 48 h load profile, which was part of the described test grid, to obtain a test case comprising 48 days with a temporal resolution of 1 h, i.e. $N_T = 960$ time steps. This causes the linear system (24), which has to be solved in every PDIPM iteration, to have the dimension $n^x + n^\lambda \approx 6.168 \cdot 10^6$.

Furthermore, we set $\alpha = 0.4$, which led to ramping constraints limiting the change of power generation of conventional power plants by 40%/h of their respective maximum power generation.

We set $\epsilon_{feas} = \epsilon_{grad} = \epsilon_{cost} = \epsilon_{comp} = 10^{-5}$ as PDIPM termination criteria and limited the number of PDIPM iterations to 300. The number of GMRES iterations was restricted to 1500. The GMRES method was allowed to restart after 300 iterations. Its relative residual tolerance rtol was determined as described in Sect. 3.3.

The ASM method was not restricted and the overlap s_{ol} was set to 3 for all tests.

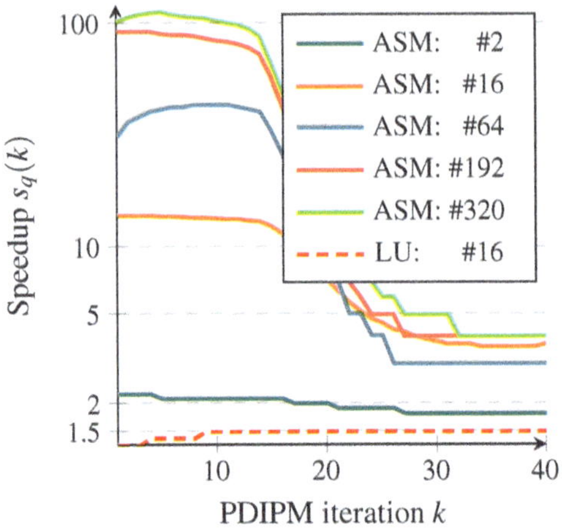

Fig. 3 Speedup up to the iteration on the x-axis for different numbers of processors

The parallel speedup on q processors for the first k PDIPM iterations is defined as $s_q(k) := \frac{T_1(k)}{T_q(k)}$. We computed the sequential reference time $T_1(k)$ by solving the linear systems (24) using a LU-factorization and adding up the times needed to solve them up to the k-th PDIPM iteration. MKL PARDISO's sequential LU-factorization algorithm was applied.

Analogously, $T_q(k)$ was obtained by adding up the times needed to solve the same linear systems, but this time iteratively, i.e. the GMRES method and the ASM were used. In Fig. 3 we plotted the results of our speedup computations for different numbers of processors. The dashed graph represents the speedup obtained when the KKT systems (24) are solved directly with MKL PARDISO's parallel implementation of the LU-factorization using 16 processors. For this amount of processors we observed the best speedup.

Besides, we scaled the y-axis logarithmically to ease drawing a comparison with the LU-solver and to emphasise the large speedups during the first iterations.[7] We observed for our proposed iterative solver, no matter the number of processors q, a clear speedup in comparison with the LU-solver. The parallelization, due to the ASM, shows its power during the first PDIPM iterations. Unfortunately, this initial speedup decreases rapidly. After 40 iterations we end up with a total speedup of five. This decrease is accompanied by an increase in GMRES iterations. To demonstrate this, we plotted in Fig. 4 the number of GMRES iterations needed to solve the KKT system at the k-th PDIPM iteration. The first thing to note is, that the number of GMRES iterations does not only change with PDIPM iterations, but also with the number of processors used to solve the KKT systems. It's likely that the increase in GMRES iterations with PDIPM iterations is caused by a change in the spectrum of the original KKT matrix A. The increase with the number of

[7]We remark that the PDIPM needed in most of the tested cases more than 40 iterations to converge, yet, it converged after at most 45 iterations.

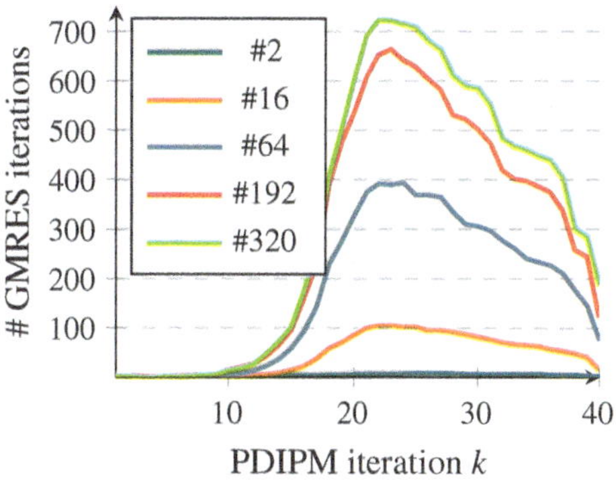

Fig. 4 The number of GMRES iterations needed to solve the KKT system corresponding to the PDIPM iteration on the x-axis

processors can be explained as follows: The more proccesors we use, the smaller the subproblems, described by A_l, become and, hence, the more intertemporal couplings are neglected, which reduces the "approximity" of M_{ASM} to A^{-1}. And the closer M_{ASM} is to A^{-1}, the more the spectrum of the preconditioned operator $\sigma(AM_{ASM})$ is clustered around 1.

Before we start to actual investigate the spectrum of the matrices in the next part of this section, we would like to summarise our findings. Even though we observed a rapid decrease in the speedup, it was possible to obtain a moderate speedup with only 16 processors and in comparison to a standard parallel LU-factorization also using 16 processors, our solver was more than a factor 2 faster.

5.2 *Eigenvalue Distribution*

Since the GMRES method converges faster if the eigenvalues of matrix A are clustered around 1, i.e. it needs less iterations, we would like to investigate how much the ASM, applied as preconditioner, improves the eigenvalues' distribution. Therefore, we calculated the eigenvalues with the largest and smallest magnitude of the KKT matrix A and the preconditioned KKT matrix AM_{ASM}, respectively. We write $\lambda^A_{max} = \max\{|\lambda|: \lambda \in \sigma(A)\}$ and $\lambda^A_{min} = \min\{|\lambda|: \lambda \in \sigma(A)\}$. Seeing that it takes very long to compute, or that it is even not possible to compute the eigenvalues for the complete $N_T = 960$ time steps, we decided to calculate them for $N_T = 96$ time steps and for each PDIPM iteration. We set the overlap to $s_{ol} = 1$ and used $q = 32$ subdomains. The results are presented in Fig. 5. The red lines depict the development of λ^A_{max} and λ^A_{min} of the KKT matrix, respectively. Thus, they form the limits for all possible magnitudes of eigenvalues corresponding to A. The latter is indicated by the red shaded area. Analogously, the blue lines and the blue area represent the spectrum of the preconditioned KKT matrix.

The spectrum of the preconditioned KKT matrix, is clustered around 1 in every interior point method iteration. Looking at the magnified part of the graph shows

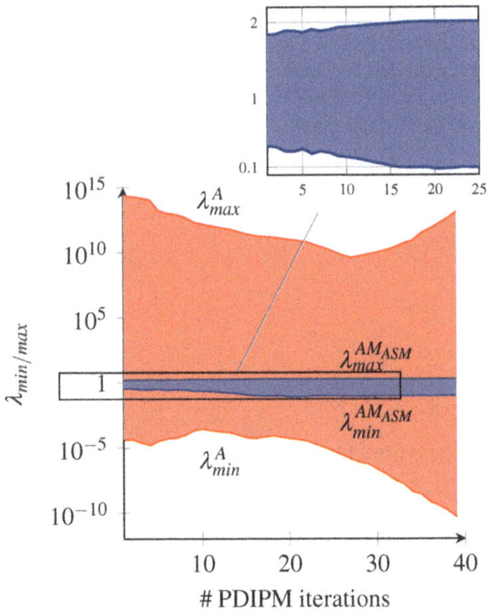

Fig. 5 Variation of the spectrum of the original KKT matrix and the preconditioned one

that $\lambda_{min}^{AM_{ASM}}$ becomes smaller between the 10-th and the 20-th iteration. This may explain the increase in GMRES iterations observed in our numerical experiments and plotted in Fig. 4.

6 Conclusion

In this work we proposed a way of solving linear systems arising from Dynamic Optimal Power Flow (DOPF) problems in parallel by means of an overlapping Schwarz domain decomposition method, namely the additive Schwarz method. It was shown how to apply this method in the context of DOPF problems and that the obtained submatrices correspond to localised formulations of a DOPF problem with additional boundary conditions. Numerical tests on one large-scale DOPF problem showed that the combination of the Generalized Minimal Residual (GMRES) method and the Additive Schwarz Method (ASM) is able to solve DOPF problems. Furthermore, we were able to show that this combination yields good parallel speedup for some of the PDIPM iterations. And, in a small example, we demonstrated that the ASM is a good preconditioner in the sense, that it clusters the eigenvalues around 1.

Unfortunately, the proposed solver had problems with some of the KKT systems arising in the large-scale DOPF problem. Therefore, it is necessary to improve it and this will be one of our main goals in future work.

Acknowledgements This work was carried out with the financial support of the German Research Foundation (DFG) within the project 242471205. Moreover, the authors acknowledge support by the state of Baden-Württemberg through bwHPC and the German Research Foundation (DFG) through grant INST 35/1134-1 FUGG.

References

1. N. Alguacil, A.J. Conejo, Multiperiod optimal power flow using benders decomposition. IEEE Trans. Power Syst. **15**(1), 196–201 (2000)
2. S. Balay, W.D. Gropp, L. Curfman McInnes, B.F. Smith, Efficient management of parallelism in object oriented numerical software libraries, in *Modern software tools in scientific computing*, ed. by E. Arge, A.M. Bruaset, H.P. Langtangen (Birkhäuser Press, Boston 1997), pp. 163–202
3. S. Bellavia, Inexact interior-point method. J. Optim. Theory Appl. **96**(1), 109–121 (1998)
4. X.-C. Cai, M. Sarkis, A restricted additive Schwarz preconditioner for general sparse linear systems. SIAM J. Sci. Comput. **21**(2), 792–797 (1999)
5. F. Capitanescu, M. Glavic, D. Ernst, L. Wehenkel, Interior-point based algorithms for the solution of optimal power flow problems. Electr. Power Syst. Res. **77**(5–6), 508–517 (2007)
6. C.Y. Chung, W. Yan, F. Liu, Decomposed predictor-corrector interior point method for dynamic optimal power flow. IEEE Trans. Power Syst. **26**(3), 1030–1039 (2011)
7. F.E. Curtis, J. Huber, O. Schenk, A. Wächter, A note on the implementation of an interior-point algorithm for nonlinear optimization with inexact step computations. Math. Program. **136**(1), 209–227 (2012)
8. P. Gerstner, V. Heuveline, M. Schick, A multilevel domain decomposition approach for solving time constrained optimal power flow problems. Preprint Ser. Eng. Math. Comput. Lab (EMCL) **4**, 1–30 (2015)
9. V. Hernandez, J.E. Roman, V. Vidal, SLEPc: a scalable and flexible toolkit for the solution of eigenvalue problems. ACM Trans. Math. Softw. **31**(3), 351–362 (2005)
10. C. Keller, N.I.M. Gould, A.J. Wathen, Constraint preconditioning for indefinite linear systems. SIAM J. Matrix Anal. Appl. **21**(4), 1300–1317 (2000)
11. A. Meister, *Numerik linearer Gleichungssysteme. Eine Einführung in moderne Verfahren. Mit MATLAB-Implementierungen von C. Vömel*, 5th revised edn. (Springer Spektrum, Heidelberg, 2015)
12. N. Meyer-Hübner, M. Suriyah, T. Leibfried, V. Slednev, V. Bertsch, W. Fichtner, P. Gerstner, M. Schick, V. Heuveline, Time constrained optimal power flow calculations on the German power grid, in *International ETG Congress 2015; Die Energiewende – Blueprints for the new energy age*, Bonn, 2015, pp. 1–7
13. N. Meyer-Hübner, F. Gielnik, M. Suriyah, T. Leibfried, Dynamic optimal power flow in AC networks with multi-terminal HVDC and energy storage, in *2016 IEEE Innovative Smart Grid Technologies – Asia (ISGT-Asia)*, Nov 2016
14. J. Nocedal, S.J. Wright, *Numerical optimization*. Springer Series in Operations Research and Financial Engineering (Springer, Berlin, 2006)
15. Y. Saad, M.H. Schultz, GMRES: a generalized minimal residual algorithm for solving nonsymmetric linear systems. SIAM J. Sci. Stat. Comput. **7**(3), 856–869 (1986)
16. M. Schönfelder, A. Eßer-Frey, M. Schick, W. Fichtner, V. Heuveline, T. Leibfried, New developments in modeling network constraints in techno-economic energy system expansion planning models. Zeitschrift für Energiewirtschaft **36**(1), 27–35 (2012)
17. H.A. Schwarz, *Ueber einen Grenzübergang durch alternirendes Verfahren*. Vierteljahrsschrift der Naturforschenden Gesellschaft in Zürich. Zürcher u. Furrer, 1870
18. B.F. Smith, P.E. Bjørstad, W.D. Gropp, *Domain Decomposition: Parallel Multilevel Methods for Elliptic Partial Differential Equations* (Cambridge University Press, Cambridge, 1996)

19. E.M. Soler, V.A. de Sousa, G.R.M. da Costa, A modified primal-dual logarithmic-barrier method for solving the optimal power flow problem with discrete and continuous control variables. Eur. J. Oper. Res. **222**(3), 616–622 (2012)
20. K. Xie, Y.H. Song, Dynamic optimal power flow by interior point methods. IEE Proc. Gener. Transm. Distrib. **148**(1), 76–84 (2001)
21. R.D. Zimmerman, C.E. Murillo-Sánchez, R.J. Thomas, Matpower: steady-state operations, planning, and analysis tools for power systems research and education. IEEE Trans. Power Syst. **26**(1), 12–19 (2011)
22. J. Zhu, *Optimization of Power System Operation*, vol. 47 (John Wiley & Sons, Hoboken, 2015)

Part II
Energy System Integration

Optimal Control of Compressor Stations in a Coupled Gas-to-Power Network

Eike Fokken, Simone Göttlich, and Oliver Kolb

Abstract We introduce a tool for simulation and optimization of gas pipeline networks coupled to power grids by gas-to-power plants. The model under consideration consists of the isentropic Euler equations to describe the gas flow coupled to the AC powerflow equations. A compressor station is installed to control the gas pressure such that certain bounds are satisfied. A numerical case study is presented that showcases effects of fast changes in power demand on gas pipelines and necessary operator actions.

Keywords Coupling of gas and power networks · Compressor stations · Optimal control

Mathematics Subject Classification (2010) 76N15, 65M08, 49J20

1 Introduction

Renewable power sources have an ever increasing share of all power sources. Though renewable energy has been developed in recent years with great success, its intermittent and unpredictable nature raises the difficulty to balance the energy production and consumption [6, 18]. A frequent proposal is to use gas turbine plants to compensate for sudden drops in power of renewable sources because these plants are relatively flexible in comparison to coal or nuclear plants. A welcome advantage of this approach is the possibility to run gas plants with fuel produced from renewable electricity via power-to-gas plants, thereby reducing or even negating carbon emissions of the gas plants. For a review of power-to-gas capability see [6].

The authors gratefully thank the BMBF project ENets (05M18VMA) for the financial support.

E. Fokken (✉) · S. Göttlich · O. Kolb
University of Mannheim, Mannheim, Germany
e-mail: fokken@uni-mannheim.de; goettlich@uni-mannheim.de; kolb@uni-mannheim.de

V. Bertsch et al. (eds.), *Advances in Energy System Optimization*,
Trends in Mathematics, https://doi.org/10.1007/978-3-030-32157-4_5

It is desirable to have a joint optimal control framework for power and gas sector of the energy system to model this compensation. So far only steady-state flow in the gas network has been considered [3, 18, 20], which may be too coarse for several applications. Therefore, we focus on an optimal control strategy for the instationary gas network model [1] coupled to a power grid [2] via compressor stations, see for example [8, 15, 17]. The mathematical foundation for the gas-to-power coupling has been recently introduced in [5], where conditions for the well-posedness have been derived and proved. This work is the first attempt to model this interaction and yields an understanding of the underlying equations. The next step will be real-world scenarios.

2 Optimal Control Problem

The gas dynamics within each pipeline of the considered gas networks are modeled by the isentropic Euler equations, supplemented with suitable coupling and boundary conditions. For the power grid, we apply the well-known powerflow equations. The coupling between gas and power networks at gas-driven power plants is modeled by (algebraic) demand-dependent gas consumption terms. To react on the demand-dependent influences on the gas network, controllable devices as compressor stations are considered within the gas network. The aim is to fulfill given state restrictions like pressure bounds whereas at the same time the entire fuel gas or power consumption of the compressor stations is to be minimized.

The network under consideration is similar to the one presented in [5] and is depicted in Fig. 1. In addition, there is now a compressor with a time-dependent control $u(t)$, which is used to satisfy pressure bounds. The optimal control problem is to minimize compressor costs while satisfying power demand and gas dynamics:

$$\begin{aligned} \min_{u(t)}\ & \text{compressor costs} \\ \text{s.t.}\ & \text{isentropic Euler equations (Sect. 2.1)} \\ & \text{gas coupling conditions (Sect. 2.2)} \\ & \text{compressor equation (Sect. 2.3)} \\ & \text{powerflow equations (Sect. 2.4)} \\ & \text{gas-power-coupling (Sect. 2.5)} \\ & \text{pressure bounds.} \end{aligned}$$

Mathematically, this is an instationary nonlinear optimization problem constrained by partial differential equations, see [9] for an overview. To solve the problem, we make use of a first-discretize-then-optimize approach and apply the interior point solver IPOPT [16]. The necessary gradient information for IPOPT, i.e., gradients with respect to all controllable devices, is efficiently computed via adjoint equations. Here, the underlying systems can be solved time-step-wise (backwards in time), where additionally the sparsity structure is exploited.

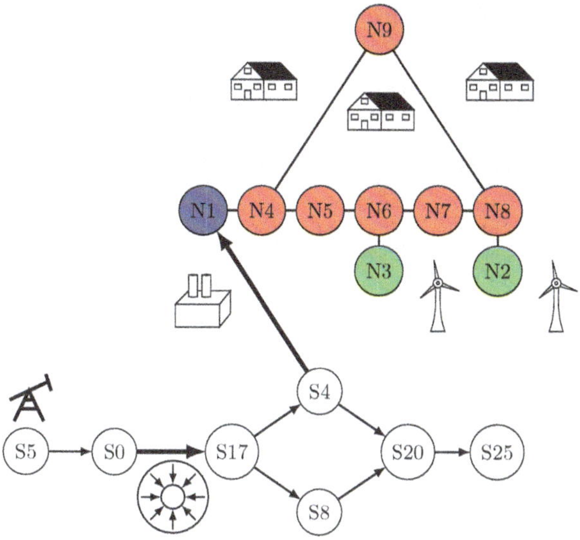

Fig. 1 Gas network connected to a power grid. Red nodes are PQ/demand nodes, green nodes are generators (PV nodes) and the blue node is the slack bus (also a generator, with gas consumption of the form $\varepsilon(P) = a_0 + a_1 P + a_2 P^2$). The circle symbol indicates a compressor station

We remark that the cost function is given in Sect. 2.3, while the bounds on the pressure are introduced as box constraints within the numerical optimization procedure in Sect. 3. Within the Sects. 2.1, 2.2, 2.3, 2.4, and 2.5, we now describe the constraints of the optimal control problem in detail and focus on the technical details.

2.1 *The Isentropic Euler Equations*

The gas network is modeled by the isentropic Euler equations (see [1, 4]), which govern gas flow in each pipeline between nodes,

$$\begin{pmatrix} \rho \\ q \end{pmatrix}_t + \begin{pmatrix} q \\ p(\rho) + \frac{q^2}{\rho} \end{pmatrix}_x = \begin{pmatrix} 0 \\ S(\rho, q) \end{pmatrix}, \tag{1}$$

where ρ is the density, $q = \rho v$ is the flow, $p(\rho) = \kappa \rho^\gamma$ is the pressure function. In our example we use $\gamma = 1$ and $\kappa = c^2 = (340\frac{\mathrm{m}}{\mathrm{s}})^2$, that is, the isothermal Euler equations with speed of sound c. The Euler equations must be met for $0 \leq x \leq l_e$

and $t \geq 0$, where l_e is the length of the pipe. Furthermore, S is a source term,

$$S(\rho, q) = -\frac{\lambda(q)}{2d_e}\frac{q|q|}{\rho},$$

where λ is the solution to the Prandtl-Colebrook formula,

$$\frac{1}{\sqrt{\lambda}} = -2\log_{10}\left(\frac{2.51}{Re(q)\sqrt{\lambda}} + \frac{k_e}{3.71d_e}\right).$$

The Reynolds number Re is given by

$$Re(q) = \frac{d_e}{\eta}q$$

with dynamic viscosity

$$\eta \;= 10^{-5}\frac{\text{kg}}{\text{ms}}.$$

The roughness k_e and diameter d_e of the pipes are all the same in our example and given by

$$k_e = 5 \cdot 10^{-4}\,\text{m},$$
$$d_e = 6 \cdot 10^{-1}\,\text{m}.$$

2.2 *Coupling at Gas Nodes*

At the nodes we use the usual Kirchhoff-type coupling conditions: The pressure is the same near the node in all pipes connected to it and the flows must add up to zero (where the sign for inflow is positive, the sign for outflow negative),

$$p_{\text{pipe}} = p_{\text{node}}, \tag{2a}$$

$$\sum_{\text{ingoing pipes}} q_{\text{pipe}} = \sum_{\text{outgoing pipes}} q_{\text{pipe}}. \tag{2b}$$

The example gas network we are using is a small part of the GasLib-40 network [10]. In Table 1 the only remaining parameter, the length l_e of each pipe, is gathered. Note that no length is given for the arc connecting S0 and S17, because only a compressor is situated between these nodes.

Table 1 Parameters of the gas network

Pipe	From	To	Length l_e [km]
P10	S4	S20	20.322
P20	S5	S0	20.635
P21	S17	S4	10.586
P22	S17	S8	10.452
P24	S8	S20	19.303
P25	S20	S25	66.037

2.3 Compressor Stations

To compensate for pressure losses in the gas network, we consider compressor stations, which are also modelled as arcs. Those arcs have (time-dependent) in- and outgoing pressure (p_{in}, p_{out}) and flux values (q_{in}, q_{out}). In general, the two separate flux values allow the modelling of fuel gas consumption of the compressor station, whereas we will consider an external power supply for the compressor and therefore have $q_{\text{in}} = q_{\text{out}}$. The power consumption is modelled as a quadratic function of the power required for the compression process. Denoting this as function $P(p_{\text{in}}(t), q_{\text{in}}(t), p_{\text{out}}(t), q_{\text{out}}(t))$, our objective function in the optimal control problem is of the form

$$\int_0^T P(p_{\text{in}}(t), q_{\text{in}}(t), p_{\text{out}}(t), q_{\text{out}}(t))dt. \tag{3}$$

Note that the power consumption does not influence the network dynamics and is therefore only of interest for the optimization procedure. For our investigations below it is sufficient to know that the consumption and therewith the costs increase if the ratio $p_{\text{out}}/p_{\text{in}}$ increases. For the details of the power consumption model, we refer to [17, Section 3.2.3]. The influence of the compressor station on the network dynamics is modelled by the control $u(t)$ of the pressure difference:

$$p_{\text{out}}(t) - p_{\text{in}}(t) = u(t).$$

2.4 Power Model

For the power grid we use the AC powerflow equations (see [7] for an introduction),

$$
\begin{aligned}
P_k &= \sum_{j=1}^{N} |V_k||V_j| \left(G_{kj}\cos(\phi_{k,j}) + B_{kj}\sin(\phi_{k,j})\right),\\
Q_k &= \sum_{k=1}^{N} |V_k||V_j| \left(G_{kj}\sin(\phi_{k,j}) - B_{kj}\cos(\phi_{k,j})\right),
\end{aligned}
\tag{4}
$$

where P_k, Q_k are real and reactive power at node k, $|V_k|$ is the voltage amplitude, ϕ_k is the phase (and $\phi_{k,j} = \phi_k - \phi_j$) and B_{kj}, G_{kj} are parameters of the transmission lines between nodes k and j or of the node k for B_{kk} and G_{kk}. Each node is either the slack bus (in our case N1; V and ϕ given), a generator bus (N2 and N3; V and P given) or a load bus (N4 through N9; P and Q given). All in all for N nodes we get $2N$ equations for $2N$ variables.

The considered power grid is taken from the example "case9" of the MATPOWER Matlab programming suite [19]. A per-unit system is used, whose base power and voltage are 100 MW and 345 kV respectively. The corresponding node and transmission line parameters are gathered in Table 2, these are the entries of the nodal admittance matrix (see [7]).

2.5 Coupling

The last ingredient is a model for converting gas to power at a gas power plant. In our example, it will be situated between the nodes S4 and N1 and convert a gas flow ε into a real power output P according to

$$
\varepsilon(P) = a_0 + a_1 P + a_2 P^2. \tag{5}
$$

The flow ε must be diverted from the pipeline network, hence the coupling condition at node S4 must be changed to

$$
\begin{aligned}
p_{\text{in}} &= p_{\text{out}},\\
q_{\text{in}} &= q_{\text{out}} + \varepsilon\,.
\end{aligned}
\tag{6}
$$

The details of simulation of such a combined network and the treatment of all arising mathematical issues can be found in [5]. We now showcase a concrete example.

Table 2 Parameters of the power grid (p.u.)

(a) Busses		
Node	G	B
N1	0.0000	−17.3611
N2	0.0000	−16.0000
N3	0.0000	−17.0648
N4	3.3074	−39.3089
N5	3.2242	−15.8409
N6	2.4371	−32.1539
N7	2.7722	−23.3032
N8	2.8047	−35.4456
N9	2.5528	−17.3382

(b) Transmission lines				
Edge	From	To	G	B
TL14	N1	N4	0.0000	17.3611
TL45	N4	N5	−1.9422	10.5107
TL56	N5	N6	−1.2820	5.5882
TL36	N3	N6	0.0000	17.0648
TL67	N6	N7	−1.1551	9.7843
TL78	N7	N8	−1.6171	13.6980
TL82	N8	N2	0.0000	16.0000
TL89	N8	N9	−1.1876	5.9751
TL94	N9	N4	−1.3652	11.6041

3 Numerical Results

3.1 Problem Setup

As already noted, we consider a small part of the GasLib-40 network from [2] consisting of 7 pipelines with a total length of 152 km. This network is extended by a compressor station and additionally connected to a power grid with 9 nodes by a gas-to-power generator. For this coupled gas-power network, we simulate a sudden increase in power demand within the power grid and study its effect on the gas network. The considered compressor station is supposed to compensate part of the pressure losses in the gas network such that a given pressure bound is satisfied all the time, while power consumption of the compressor is minimized.

To complete the problem description, the following initial and boundary conditions are given:

- $P(t)$ and $Q(t)$ at load nodes,
- $P(t)$ and $V(t)$ at generator nodes,
- $V(t)$ and $\phi(t)$ at the slack bus,
- $p(t)$ at S5,
- $q(t)$ at S25,
- $p(x, 0)$, $q(x, 0)$ for all pipelines.

Table 3 Initial conditions of the power grid (p.u.)

Node	P	Q	$\|V\|$	ϕ
N1	–	–	1	0
N2	163	–	1	–
N3	85	–	1	–
N4	0	0	–	–
N5	−90	−30	–	–
N6	0	0	–	–
N7	−100	−35	–	–
N8	0	0	–	–
N9	−125	−50	–	–

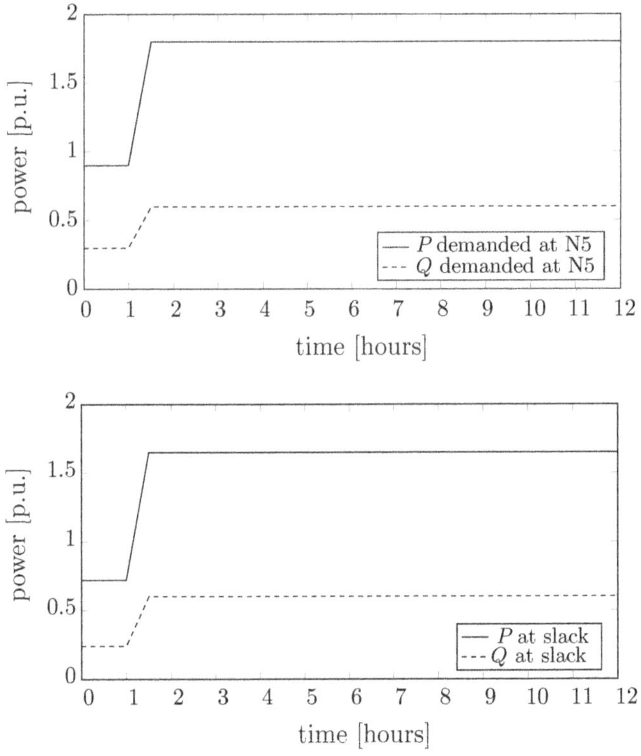

Fig. 2 Power and reactive power at demand node N5 (above) and the slack bus (below)

More precisely, the initial conditions for the power network are given in Table 3. These remain constant over time except for the power demand at node N5, which changes linearly between $t = 1\,\text{h}$ and $t = 1.5\,\text{h}$ from 0.9 p.u. to 1.8 p.u. for the real power and from 0.3 p.u. to 0.6 p.u. for the reactive power, see also Fig. 2.

For the gas network the incoming pressure at S5 is fixed at 60bar, the outflow at S25 is fixed at $q = 100\frac{\text{m}^3}{\text{s}} \cdot \frac{\rho_0}{A_e}$, where $\rho_0 = 0.785\frac{\text{kg}}{\text{m}^3}$ and $A_e = \pi\frac{d_e^2}{4}$. The fuel consumption parameters we use in equation (5) are given by $a_0 = 2$, $a_1 = 5$,

$a_2 = 10$. Since the data for the considered gas and power networks are taken from different sources, the parameters of the gas-to-power generator are chosen in such a way that a significant influence is caused.

Further, the pressure at S25 is supposed to satisfy a lower pressure bound of 41bar, i.e.,

$$p_{S25}(t) \geq 41\text{bar}$$

for all times t, where we consider a time horizon of $T = 12\,\text{h}$.

As we will see below, the compressor station between nodes S0 and S17 will have to run at a certain time, i.e., $u(t) > 0$, to keep this pressure bound. In general, a solution to the described optimal control problem consists of the control $u(t)$ and the entire network state for all times t, i.e.,

- $P(t)$, $Q(t)$, $V(t)$, $\phi(t)$ for all nodes in the power grid,
- $p(x, t)$, $q(x, t)$ for all pipelines in the gas network,

fulfilling the given model equations and the pressure constraint.

3.2 *Discretization and Optimization Schemes*

To solve the described optimal control problem, we follow a first-discretize-then-optimize approach. The model equations of the power grid only require a discretization in time, which means that the given boundary conditions and the powerflow equations (4) hold for discrete times $t_j = j\Delta t$ with $\Delta t = 15\,\text{min}$ and $j \in \{0, \ldots, 48\}$ in our scenario. The discretization of the isentropic Euler equations within the pipelines of the gas network additionally requires a spatial grid (here with grid sizes $\Delta x_e \approx 1$ km) and an appropriate discretization scheme. Here we apply an implicit box scheme [14], which allows the application of large time steps as $\Delta t = 15\,\text{min}$ for the considered spatial grid. Considering the isentropic Euler equations as a system of balance laws of the form

$$y_t + f(y)_x = g(y),$$

the applied scheme reads

$$\begin{aligned}\frac{Y_{j-1}^{n+1} + Y_j^{n+1}}{2} =& \frac{Y_{j-1}^n + Y_j^n}{2} \\ &- \frac{\Delta t}{\Delta x}\left(f(Y_j^{n+1}) - f(Y_{j-1}^{n+1})\right) \\ &+ \Delta t \frac{g(Y_j^{n+1}) + g(Y_{j-1}^{n+1})}{2}.\end{aligned}$$

The numerical approximation of the balance law is thought in the following sense:

$$Y_j^n \approx y(x,t) \quad \text{for} \quad x \in \left[(j-\tfrac{1}{2})\Delta x, (j+\tfrac{1}{2})\Delta x\right),$$
$$t \in \left[n\Delta t, (n+1)\Delta t\right).$$

Together with the algebraic equations modelling the compressor station and the coupling and boundary conditions, the discretization process results in a system of nonlinear equations for all state variables of the coupled gas-power network. For simulation purposes, the entire discretized system is solved with Newton's method. Note that the system can be solved time-step per time-step and that we exploit the sparsity structure of the underlying Jacobian matrices.

So far we can only compute the state of the considered gas-power network (for a given compressor control $u(t)$) and evaluate quantities of interest like the power consumption of the compressor station or the pressure constraint within the time horizon of the simulation. For the given time discretization, the compressor costs (3) are approximated by the trapezoidal rule and formally contained in $J(u, y(u))$ below. Next, we want to solve the (discretized) optimal control problem, i.e., to find control values $u(t_j)$ such that the pressure constraint is satisfied, while the power consumption of the compressor is minimized. For this purpose, we apply the interior point optimization code IPOPT [16] to the (reduced) discretized optimal control problem. In addition to our simulation procedure to evaluate quantities of interest for a given control, IPOPT further requires gradient information of those quantities with respect to the control. Based on the considered discretization, such information can be efficiently computed by an adjoint approach, which we briefly describe in the following. Therefore, we consider the discretized optimal control problem in the following abstract form:

$$\begin{aligned} \min\ & J(u, y(u)) \\ \text{s.t.}\ & u \in \mathbb{R}^{Nu},\ \ y \in \mathbb{R}^{Ny}, \\ & \text{model equations: } E(u, y(u)) = 0, \\ & \text{constraints: } g(u, y) \geq 0. \end{aligned}$$

The vector u contains all control variables (here the compressor control $u(t_j)$ for all times $t_j = j\Delta t \in [0, 24]$) and y contains all state variables of the coupled gas-power network of all time steps t_j. The mapping $u \to y(u)$ is implicitly given by our simulation procedure. Thus, IPOPT does not have to care about the model equations formally summarized in $E(u, y)$, but only about the further constraints $g(u, y)$ and minimizing the objective function $J(u, y)$. Accordingly, we need to provide total derivatives of J and g with respect to u.

In the following, we consider the computation of these derivatives only for the objective function, since the procedure is identical for the constraints, and we follow the description given in [12]. First of all, the chain rule yields

$$\frac{d}{du}J(u,y(u)) = \frac{\partial}{\partial u}J(u,y(u)) + \frac{\partial}{\partial y}J(u,y(u))\frac{d}{du}y(u). \tag{7}$$

While the partial derivatives of J with respect to u and y can be directly computed, the derivatives of the states y with respect to the control u are only implicitly given. Differentiating the model equations $E(u, y(u)) = 0$ yields

$$0 = \frac{d}{du}E(u,y(u)) = \frac{\partial}{\partial u}E(u,y(u)) + \frac{\partial}{\partial y}E(u,y(u))\frac{d}{du}y(u)$$

and therewith (formally)

$$\frac{d}{du}y(u) = -\left(\frac{\partial}{\partial y}E(u,y(u))\right)^{-1}\frac{\partial}{\partial u}E(u,y(u)). \tag{8}$$

Even though the partial derivatives on the right-hand-side of (8) can be directly computed, one would have to solve Nu systems of linear equations here. Instead of that, we insert (8) into (7) and get

$$\begin{aligned}\frac{d}{du}J(u,y(u)) &= \frac{\partial}{\partial u}J(u,y(u)) \\ &\quad \underbrace{-\frac{\partial}{\partial y}J(u,y(u))\left(\frac{\partial}{\partial y}E(u,y(u))\right)^{-1}}_{=\xi^T}\frac{\partial}{\partial u}E(u,y(u)).\end{aligned}$$

With the so-called adjoint state ξ as the solution of the adjoint equation

$$\left(\frac{\partial}{\partial y}E(u,y(u))\right)^T \xi = -\left(\frac{\partial}{\partial y}J(u,y(u))\right)^T, \tag{9}$$

we finally have

$$\frac{d}{du}J(u,y(u)) = \frac{\partial}{\partial u}J(u,y(u)) + \xi^T\frac{\partial}{\partial u}E(u,y(u)). \tag{10}$$

It is the fact that (9) is a single linear system and has a special structure, which can be easily exploited (see for instance [11, 13]), which makes the computation of derivatives via the presented adjoint approach very efficient. Nevertheless note that for given control variables u one still has to solve the model equations to get $y(u)$, before one may compute the gradient information via (9) and (10).

3.3 Results

We first discuss the simulation with inactive compressor. In the course of the simulation, due to the increase in power demand at node N5, the power demand at the slack bus rises as well and leads to increased fuel demand at node S4. This increases the inflow into the gas network, as can be seen in Fig. 3. Also the pressure in the final node S25 decreases and violates the pressure bound after approximately 4 h, see Fig. 4.

After the optimization procedure, the compressor station compensates part of the pressure losses in the gas network such that the pressure bound is satisfied all the time. Since the power consumption of the compressor station is minimized within the optimization, the pressure constraint is active after roughly 4 h (see again Fig. 4), i.e., the compressor station applies as little as possible energy.

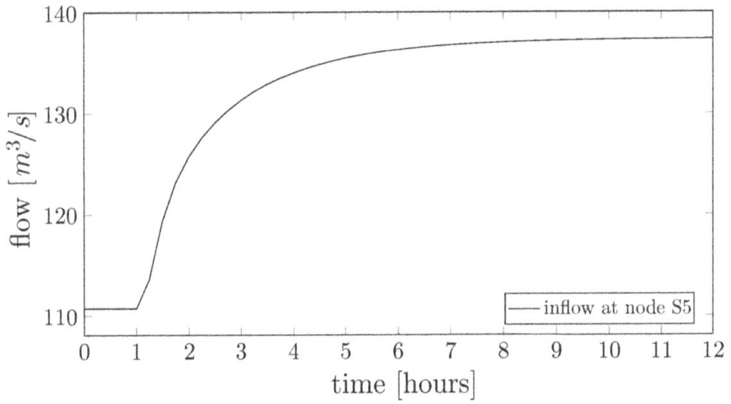

Fig. 3 Inflow at the node S5

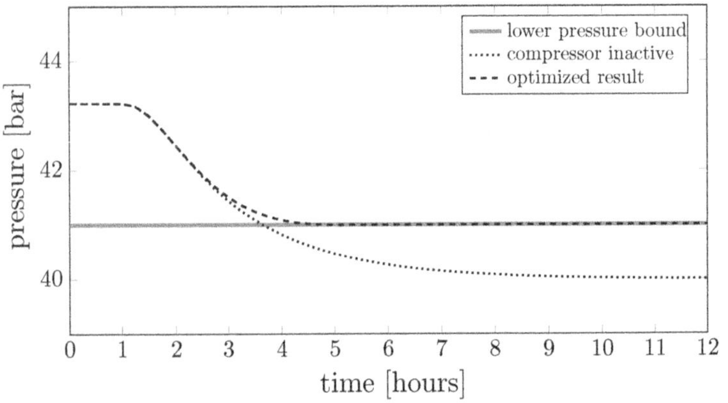

Fig. 4 Pressure at the node S25

4 Conclusion

The proposed optimization model allows to predict pressure transgressions within a coupled gas-to-power framework. Simulation and optimization tasks are efficiently solved by exploiting the underlying nonlinear problem structure while keeping the full transient regime. This makes it possible to track bounds much more accurately than with a steady state model, thereby achieving lower costs.

References

1. M. Banda, M. Herty, A. Klar, Gas flow in pipeline networks. Netw. Heterogeneous Media **1**(1), 41–56 (2006). https://doi.org/10.3934/nhm.2006.1.41
2. D. Bienstock, *Electrical Transmission System Cascades and Vulnerability* (Society for Industrial and Applied Mathematics, Philadelphia, 2015). https://doi.org/10.1137/1.9781611974164
3. M. Chertkov, S. Backhaus, V. Lebedev, Cascading of fluctuations in interdependent energy infrastructures: gas-grid coupling. Appl. Energy **160**, 541–551 (2015). https://doi.org/10.1016/j.apenergy.2015.09.085
4. R. Colombo, M. Garavello, On the cauchy problem for the p-system at a junction. SIAM J. Math. Anal. **39**(5), 1456–1471 (2008). https://doi.org/10.1137/060665841
5. E. Fokken, S. Goöttlich, O. Kolb, Modeling and simulation of gas networks coupled to power grids. arXiv e-prints (2018). arXiv: 1812.11438 [math.NA]
6. G. Gahleitner, Hydrogen from renewable electricity: an international review of power-to-gas pilot plants for stationary applications. Int. J. Hydrog. Energy **38**(5), 2039–2061 (2013). https://doi.org/10.1016/j.ijhydene.2012.12.010
7. J.J. Grainger, W.D. Stevenson, G.W. Chang, *Power System Analysis*. McGraw-Hill Series in Electrical and Computer Engineering: Power and Energy (McGraw-Hill Education, 2016). ISBN: 9781259 008351
8. M. Herty, Modeling, simulation and optimization of gas networks with compressors. Netw. Heterogeneous Media **2**(1), 81–97 (2007). https://doi.org/10.3934/nhm.2007.2.81
9. M. Hinze et al., *Optimization with PDE Constraints*. Mathematical Modelling: Theory and Applications, vol. 23 (Springer, New York, 2009), pp. xii+270. ISBN: 978-1-4020-8838-4
10. J. Humpola et al., GasLib – A Library of Gas Network Instances. Tech. rep. Sept. 2017. http://www.optimization-online.org/DB_HTML/2015/11/5216.html
11. O. Kolb, Simulation and Optimization of Gas and Water Supply Networks. Ph.D., thesis. TU Darmstadt, 2011
12. O. Kolb, S. Göttlich, A continuous buffer allocation model using stochastic processes. Eur. J. Oper. Res. **242**(3), 865–874 (2015). https://doi.org/10.1016/j.ejor.2014.10.065
13. O. Kolb, J. Lang, Simulation and continuous optimization, in *Mathematical optimization of water networks*. International Series of Numerical Mathematics, vol. 162 (Birkhäuser/Springer Basel AG, 2012), pp. 17–33
14. O. Kolb, J. Lang, P. Bales, An implicit box scheme for subsonic compressible flow with dissipative source term. Numer. Algorithms **53**(2), 293–307 (2010). https://doi.org/10.1007/s11075-009-9287-y
15. T. Mak et al., Dynamic compressor optimization in natural gas pipeline systems. INFORMS J. Comput. (2019, in press)
16. A. Wächter, L.T. Biegler, On the implementation of an interior-point filter line-search algorithm for large-scale nonlinear programming. Math. Program. **106**(1), 25–57 (2006). https://doi.org/10.1007/s10107-0040559-y

17. T. Walther, B. Hiller, Modelling compressor stations in gas networks. ZIB Report 17–67. Zuse Institute Berlin, 2017. https://opus4.kobv.de/opus4-trr154/frontdoor/index/index/docId/203
18. Q. Zeng et al., Steady-state analysis of the integrated natural gas and electric power system with bi-directional energy conversion. Appl. Energy **184.C**, 1483–1492 (2016)
19. R. Zimmerman, C. Murillo-Sanchez, R. Thomas, MATPOWER: steadystate operations, planning, and analysis tools for power systems research and education. IEEE Trans. Power Syst. **26**(1), 12–19 (2011). https://doi.org/10.1109/TPWRS.2010.2051168
20. A. Zlotnik et al., Control policies for operational coordination of electric power and natural gas transmission systems, in *2016 American Control Conference (ACC)*, July 2016, pp. 7478–7483. https://doi.org/10.1109/ACC.2016.7526854

Utilising Distributed Flexibilities in the European Transmission Grid

Manuel Ruppert, Viktor Slednev, Rafael Finck, Armin Ardone, and Wolf Fichtner

Abstract In this paper, we investigate the effect of distributed flexibilities on the operation of the transmission grid. The flexibilities considered are heat pumps, electric vehicles, battery energy storage systems and flexible renewable generation. For this purpose, we develop a two-stage approach of first determining an optimal electricity market solution considering the optimal dispatch of each generation element and flexibility. In the second step we determine the required dispatch adjustments due to transmission grid constraints and investigate the effect of integrating battery energy storage systems into the adjustable generators to solve congestions. In our case study, we investigate the central European transmission grid for a scenario based on the Distributed Generation scenario of the Ten-Year Network Development Plan for the year 2030. Integrating distributed flexibilities leads to a strong increase in the security of supply, while the overall effect on the generation adjustment is small. A comparison of the results for an AC and DC formulation shows that both approaches differ significantly in individual cases.

Keywords AC optimal power flow · Battery energy storage systems · DC optimal power flow · Distributed flexibilities · Distributed generation · Flexible demand · Transmission grid

1 Introduction

Today's change in the electricity system is driven by the decarbonisation initiative developed to meet emission reduction targets defined in international agreements in order to limit the global temperature increase. To reduce the carbon intensity of the electricity generation, an increasing amount of electricity generation from renewable sources (RES-E) is being installed throughout Europe. The two most

M. Ruppert (✉) · V. Slednev · R. Finck · A. Ardone · W. Fichtner
Chair of Energy Economics, Institute for Industrial Production, Karlsruhe Institute of Technology (KIT), Karlsruhe, Germany
e-mail: manuel.ruppert@kit.edu

V. Bertsch et al. (eds.), *Advances in Energy System Optimization*, Trends in Mathematics, https://doi.org/10.1007/978-3-030-32157-4_6

significant sources are wind and solar, both characterised by volatility and their spatially distributed generation potential [1]. On the demand side, decarbonisation efforts have begun to lead to an increased electrification in the heat and transportation sectors. In some countries, rapid market growth of such elements which can be categorised as distributed flexibilities can be already be observed today. Future energy scenarios hint towards a further increase of these developments until 2030. With potentially high concurrencies of individual, decentralised demand patterns, uncontrolled or market-driven operation of these flexibilities will increase local as well as aggregated consumptions peaks in the power sector.

These changes pose significant challenges to the operation of the European transmission grid. First, balancing supply and demand will require additional flexibilities in a system dominated by distributed, intermittent RES-E and new sources of electrical demand. Second, RES-E at a specific point in time will be largely determined by the predominant meteorological conditions with high spatial concentration, increasing average distance between generation and consumption and thus increase the stress on the grid infrastructure. Lastly, market requirements to create a cost-minimal dispatch might contradict those necessary to enable a congestion-free transmission grid.

In the past, extensive research has been performed on different flexibility types concerning the technologies, their modelling and their contribution to renewable integration. A comprehensive overview on the current developments in electrical energy storage technologies and their potential for application in power system operation is given in [2] and [3]. Current reviews of different technological options can be found for power-to-heat applications [4], residential photovoltaic-battery energy storage (PV-BESS) [5], heat pumps (HP) in smart grids [6], and battery electric vehicles (BEV) [7]. Most literature focuses on the local balancing of demand and supply, either on a single household level or a community. This neglects the implications on the grid level or focuses on the regional effects in micro grids while including only a small set of technological options.

There are numerous models, which take into account flexibilities and simplified transmission grid constraints. In most cases, these models have been developed to analyse electricity markets, e.g. for investment decision support or operation decision support. In most studies liberalised markets in the United States or Europe are analysed and grid constraints are represented in a linearised way or by import/export constraints between market zones. Hence, most techno-economic models only take these constraints into account.

Few models, which consider flexibilities in a grid context take into account the full AC formulation of transmission grid constraints. Besides the group of studies and models, which consider a linearised form of grid constraints, some models can integrate full AC constraints. However, they are limited to either a very short time horizon or their focus lies on transient technical aspects like fault level detection, transient stability, harmonic analysis, reliability and power flow. Another group of models takes into account many flexibilities and AC restrictions but only on a distribution grid level, thus limiting system size and thereby complexity significantly [8]. A comprehensive overview on modelling approaches and their consideration of the grid can be found in [9].

Another area where transmission constraints are explicitly considered is the co-optimisation of transmission and generation capacities. The complexity of the resulting problem forces modellers to apply linearised transmission grid constraints or reduce the number of modelled flexibilities [10]. Furthermore, such models are often only validated on small scale or test grids, making it difficult to assess their suitability for large real-world power grids [11]. In most cases, the focus lies on investment decisions rather than on utilising operational flexibility in future transmission systems. [12] provide an overview of requirements and approaches to address this problem.

In this paper, we present a modelling framework to investigate the effect of distributed flexibilities on the transmission grid operation and investigate the benefit of utilising distributed flexibilities to reduce grid congestion while taking into account the full AC model of the transmission grid.

2 Modelling the Interaction of Market and Grid Operation

In order to model today's interaction between the electricity market and transmission grid operation, we developed a two-step approach to model market and grid operation. In the first step, we use a linear optimisation approach to determine the minimal-cost, copperplate-based results of the interconnected electricity market. In the second step, we determine the required dispatch adjustments due to congestion in the transmission grid using a multi-period nonlinear optimisation model. In the following section, the general modelling of the electricity market is described, followed by the integration of distributed flexibilities and the application of the method for modelling the transmission grid operation.

2.1 *Electricity Market*

To model the interconnected electricity market consisting of multiple market areas, we use a linear optimisation approach with the objective function of minimal total annual system cost under the assumptions of perfect foresight and perfect competition.[1] The system cost consist of the aggregated variable cost of electricity

[1] A representation of the electricity market using an optimisation approach with cost minimisation objective function and perfect foresight does not necessarily allow for a realistic representation of market prices, as scarcitiy rents cannot be considered. As the market model is first and foremost utilised to obtain dispatch results for the transmission grid simulation in our application, using the described approach is a reasonable simplification for this purpose.

production C_g for the electricity generation $p_{g,t}$ of generator g in time step t and the cost of load shedding C_{LS} for the amount of load shedding $p_{d,t}^{LS}$ of demand d in time step t.

$$\min_{p_{g,t},p_{d,t}} \sum_{g\in G,t\in T} C_g * p_{g,t} + \sum_{d\in D,t\in T} C_{LS} * p_{d,t}^{LS}$$

In every market area m, the electricity demand $P_{d,t}$ in every time step must be balanced out against generation, load shedding (LS) and interconnector flows into the market area $p_{c,t}$.

$$\sum_{g\in G^M} p_{g,t} + \sum_{d\in D^M} p_{d,t}^{LS} + \sum_{c\in IC^M} p_{c,t} = \sum_{d\in D^M} P_{d,t} \forall\, m \in M, t \in T$$

The bounds of each generator are determined by the minimum capacity $P_{g,t}^{min}$ and maximum capacity $P_{g,t}^{max}$ of the respective generating unit. While the boundaries of conventional generation are determined by the technical parameters of the generator, the boundaries of RES-E are determined by the available generation due to the intermittent nature. Available interconnection flows between market areas are restricted by the available exchange capacity P_c^{min}, P_c^{max}. In order to ensure feasibility of the problem, LS up to total demand can be performed at each time step.

$$P_{g,t}^{min} \le p_{g,t} \le P_{g,t}^{max}\ \forall g \in G, t \in T$$

$$P_c^{min} \le p_{c,t} \le P_c^{max}\ \forall c \in IC, t \in T$$

$$0 \le p_{d,t}^{LS} \le P_{d,t}\ \forall d \in D, t \in T$$

2.2 Modelling Flexibilities

In the following, we apply a set of general equations to model various types of time-coupled flexibilities, ranging from different power storage technologies to consumption and renewable flexibilities. Ignoring self-discharging losses, the energy level $v_{s,t}$ of a storage $s \in S$ at time step $t \in T = \{1, \ldots NT\}$ equals its previous level, the external power inflow $\zeta_{s,t}^{in}$ or outflow $\zeta_{s,t}^{out}$ and the sum of charged and discharged power provided by connected generators $p_{g,t} \in G^S \subseteq G$:

$$v_{s,t} = v_{s,t-1} + \sum_{g\in G^{inS}} p_{g,t} * \eta_g - \sum_{g\in G^{outS}} p_{g,t}/\eta_g + \zeta_{s,t}^{in} - \zeta_{s,t}^{out}$$

$$\forall s \in S, t \in T$$

For a simpler notation, the set of storage-connected generators G^S is split such that G^{outS} denotes those generators supplying power to the electricity system by discharging a storage and G^{inS} vice versa (charging of a storage). The charging and discharging efficiency is denoted by η_g. Furthermore, all variables might be restricted to lower and upper bounds and the starting condition for the storage is either a coupling of the first and last time steps or the definition of an initial storage level:

$$V_{s,t}^{min} \leq v_{s,t} \leq V_{s,t}^{max} \; \forall s \in S, t \in T$$

$$P_{g,t}^{min} \leq p_{g,t} \leq P_{g,t}^{max} \; \forall g \in G, t \in T$$

$$Z_t^{min} \leq \zeta_{s,t} \leq Z_t^{max} \; \forall s \in S, t \in T$$

$$v_{s,o} = v_{s,NT} \; \vee \; v_{s,o} = V_s^{start} \; \forall s \in S$$

2.3 Storage Technologies

Modelling the flexibility provided by actual storage technologies $S^S \subseteq S$ based on the equations shown above is straightforward and could easily be applied to hydro storages and batteries. Pure hydro pump storages and large-scale battery systems allow a rather simple definition of variable bounds. Storage volumes and generator capacities are non-negative and time independent, so that the equations for the energy level and power output can be reformulated.

$$0 \leq v_{s,t} \leq V_s^{max} \; \forall s \in S^S, t \in T$$

$$0 \leq p_{g,t} \leq P_g^{max} \; \forall g \in G^{inS^s}, t \in T$$

$$0 \leq p_{g,t} \leq P_g^{max} \; \forall g \in G^{outS^s}, t \in T$$

While large-scale battery systems allow to ignore external power inflow and outflow, which do not result from indirectly connected generators or consumption, this is only true for non-cascading pump storage systems. However, we will ignore the case of cascading systems in the following and refer to the description of its modelling [13]. Including a time series of natural inflow and the restriction of a potential minimal outflow, allows for modelling mixed-hydro pump storages and/or pure seasonal storages, with an empty set G^{inS} in the last case.[2]

[2]In case of missing data for modelling seasonal storages, such as the actual volume or the inflow time series, it might be modelled as described in Sect. 2.5.

Concerning small scale batteries, such as BESS combined with a PV system, the mathematical modelling is identical to large-scale battery systems, as long as no further restrictions apply to the battery utilisation. In practical application, however, regulators might apply specific restrictions for the operation of such systems. In Germany, for example, a widely used support scheme for PV-BESS is limiting the grid feed-in up to 50% of the PV capacity.[3] Given a demand $P_{d,t}$ and a maximum grid feed-in P_g^{lim}, these restriction might be easily implemented:

$$\sum_{g \in G^{PV^s}} (p_{g,t} - P_{g,t}^{lim}) + \sum_{g \in G^{out S^s}} p_{g,t} \leq \sum_{d \in D^s} P_{d,t}$$

$$\forall s \in S^{PVBESS}, t \in T$$

2.4 Demand Flexibilities

In general, consumption processes with the ability to store energy can be directly modelled as single storages. Under the assumption that a certain share of a specific load might be shifted within a certain time range, the storage restrictions, however, might be applied to a broader range of demand flexibilities or for modelling the flexible potential of aggregated loads. Considering our target to model the impact of demand flexibilities at high and extra high voltage levels, we focus on modelling the flexibility for large-scale consumption process or of aggregated loads as storages $S^L \subseteq S$. When looking at the load shifting potential of an EV fleet, aggregation allows to neglect technical restrictions of single units and to derive the load shifting potential from the specific load profile [14]. This finding is supported by [15] where an aggregated modelling of BEV as flexible loads led to almost the same results in a unit commitment problem as their individual representation, after the BEVs where clustered accordingly.

Therefore, we model the demand flexibility of utility scale HP and the aggregation of numerous BEV and household heat pumps (HH-HP) by defining certain share α of a specific load $P_{d,t}$ as flexible $P_{d,t}^{flex} = \alpha^{flex} * P_{d,t}$ and adjust the general equations based on the load profile in a way such that $P_{d,t}^{flex}$ might be shifted within a specific time interval. Given a segmentation of the time interval $T = \{TT_1, \ldots, TT_M\}$ into M subsets, the storage equation might be reduced by

[3]KfW-Kredit 275-Erneuerbare Energien-Speicher, from https://www.kfw.de/inlandsfoerderung/Unternehmen/EnergieUmwelt/F%C3%B6rderprodukte/Erneuerbare-Energien-%E2%80%93-Speicher-(275)

neglecting the variables for storage inflow and discharge:

$$v_{s,t} = v_{s,t-1} + \sum_{g \in G^{inS^s}} p_{g,t} * \eta_g - \zeta_{s,t}^{out} \forall s \in S^L, t \in T \backslash 1$$

By restricting the storage outflow to the flexible demand profile

$$Z_{s,t}^{min} \leq \zeta_{s,t}^{out} \leq Z_{s,t}^{max} = \sum_{d \in D^s} P_{d,t}^{flex} \forall s \in S^L, t \in T$$

We define that the flexible demand either directly translates to a system load or is stored. In case that load shedding is not allowed, $Z_{s,t}^{min}$ equals $Z_{s,t}^{max}$. Assuming that the energy of a flexible demand profile has to be consumed within a time interval TT_m, the bounds of the storage volume are defined by the integral of the profile over TT_m and equal zeros in the last time step.

$$V_{s,t}^{max} = -V_{s,t}^{min} = \begin{cases} 0 & if \ t = \max\{TT_m\} \\ \sum_{t \in TT_m} \sum_{d \in D^s} P_{d,t}^{flex} & else \end{cases}$$

$$\forall s \in S^L, m = 1 \dots M, t \in T$$

Concerning the bounds for the storage discharge, the minimal or maximal values of the profile time series within the certain time interval TT_m or the whole time horizon can be used.

$$P_{d,t}^{max} = \min\{ \max_{t \in TT_m} \alpha^+ * P_{d,t}^{flex}, \max_{t \in T} P_{d,t}^{flex} \}$$

$$P_{d,t}^{min} = \max\{ \min_{t \in TT_m} \alpha^- * P_{d,t}^{flex}, 0\}$$

$$\forall d \in D^s, m = 1 \dots M, t \in T$$

In order to model the specific flexible demand technologies, only the definition of the time intervals as well as α^{flex}, α^+, α^- have to be adjusted, where α^+ and α^- are scaling parameters within the range of zero and infinity.

2.5 *Renewable Flexibilities*

Given a generation profile, a flexible operation of RES technologies, which depend on a storable resource, such as hydro- or bioenergy, might be modelled analogously to the case of flexible demand. This might be useful in cases where the actual storage parameters are unknown or for modelling virtual power plants, comprising multiple small units. The general idea is to model the flexibility by enabling a shift of the initial generation profile within a certain time interval. If a certain share α of the renewable generation is based on stored energy, we first split the generation variable into a flexible and inflexible part:

$$P_{g,t}^{min} \leq p_{g,t}^{inflex} + p_{g,t}^{flex} \leq P_{g,t}^{max}$$

$$p_{g,t}^{flex} \leq \alpha^{flex} * P_{g,t}^{max} = P_{g,t}^{flex}$$

Just like in the case of flexible demand, we assume that the flexible generation might be shifted within a specific time interval $T = \{TT_1, \ldots, TT_M\}$. In consequence, the flexible operation might be expressed by modifying the general storage equation:

$$v_{s,t} = v_{s,t-1} - \sum_{g \in G^{outS^s}} p_{g,t}^{flex} / \eta_g + \zeta_{s,t}^{in}$$

$$\forall s \in S^R, t \in T \backslash 1$$

S^R defines the set of storages used to model the generation shift flexibility of renewables. By restricting the storage inflow to the flexible generation profile

$$Z_{s,t}^{min} \leq \zeta_{s,t}^{in} \leq Z_t^{max} = \sum_{g \in G^{outS^s}} P_{g,t}^{flex}$$

We define that the flexible renewable generation is either directly feed-in or stored. In case that dumping the flexible renewable generation is not allowed, $Z_{s,t}^{min}$ equals $Z_{s,t}^{max}$. Assuming that the energy of a flexible renewable generation profile has to be fed in within a time interval TT_m, the bounds of the storage volume are defined by the integral of the profile over TT_m and equal zero in the last time step.

$$V_{s,t}^{max} = -V_{s,t}^{min} = \begin{cases} 0 & if \ t = \max\{TT_m\} \\ \sum_{t \in TT_m} \sum_{g \in G^{outS^s}} P_{g,t}^{flex} & else \end{cases}$$

$$\forall s \in S^R, m = 1 \ldots M, t \in T$$

Concerning the bounds for the storage discharge, the minimal or maximal values of the profile time series within the certain time interval TT_m or the whole time horizon can be used.

$$P_{g,t}^{max} = \min\{ \max_{t \in TT_m} \alpha^+ * P_{g,t}^{flex} , \max_{t \in T} P_{g,t}^{flex} \}$$

$$P_{g,t}^{min} = \max\{ \min_{t \in TT_m} \alpha^- * P_{g,t}^{flex} , 0\}$$

$$\forall g \in G^{outS^s}, t \in T$$

In order to model the specific flexible renewable technologies, only the definition of the time intervals and α^{flex}, α^{+}, α^{-} has to be adjusted.

2.6 Grid Operation

Based on the dispatch result for each generation unit and time step from the market model, we determine the necessary dispatch adjustment due to grid constraints in the second step using a nonlinear optimisation approach. As a conventional cost minimising objective function under consideration of grid constraints would lead to a nodal pricing result, which would discard the dispatch determined by the market, we establish the minimum amount of dispatch adjustment in the entire network as the optimisation objective and evaluation figure to assess the efficiency in the transmission grid. Therefore, the generalised objective function can be formulated as follows:

$$\min_{\Delta p_{g,t},\Delta p_{d,t}^{LS}} \sum_{g\in G,t\in T} |\Delta p_{g,t}| + \sum_{d\in D} \Delta p_{d,t}^{LS}$$

Positive and negative deviation of generation units from the base solution $\Delta P_{g,t}$ are considered equivalently towards the objective, while grid-induced load shedding $\Delta P_{d,t}^{LS}$ is unidirectional by definition.

We solve the resulting optimisation problem with a multi-period nonlinear formulation of the ACOPF utilising an augmented Lagrangian formulation for coupling of time-dependent storage units as described in [16]. In order to eliminate the possibility of results representing significantly suboptimal solutions caused by converging into local optima due to the nonlinear nature of the problem, we benchmark our results with a DCOPF formulation, as described in [17].

3 Case Study

In the following, we apply the presented modelling approach on a scenario derived from the scenario "Distributed Generation" (DG) of the European Network of Transmission System Operators for Electricity (ENTSO-E) Ten-Year Network Development Plan (TYNDP) 2016. In our study, we restrict the scope of both electricity market and grid to the central European countries and their transmission networks, thus not considering the effect of flows from further interconnected countries. The considered area contains the countries France, Belgium, Netherlands, Switzerland, Austria, Czech Republic, Poland, Germany and Luxembourg. These countries form individual market areas, with the exception of Germany and Luxembourg, which form a common market area.

In addition to the scenario data from the DG scenario, the closely related high renewables scenario C of the german network development plan 2030 was used for Germany due to the fact that this source includes data with a higher level of detail. Based on the projected national RES capacity development of the scenario an optimal allocation planning model as described in [18] was run in the first step for regionalisation. In detail, a cost optimal expansion planning with a high spatial resolution until 2050 is performed, considering national and regional lower and upper bounds for the development of single renewable technologies. Due to the longer optimization horizon, exceeding the forecast horizon of the underlying scenarios, a minimal RES-E share on the demand is defined in the later years, such that an 80% renewable share is achieved in Europe until 2050.

In the second step, the capacity adequacy is analysed based on a Monte-Carlo simulation of the generation availability. Based on an integrated European approach, accounting for a cross-border balancing, and the availability of a flexible demand shift (BEV and heap pumps in households and district heating) and flexible renewables (hydro storage and biomass) we computed the needed dispatchable capacity.[4] Afterwards we utilised results obtained from a generation expansion planning problem of the European power system in order to meet the calculated capacity demand for a secure operation of the system [19].

In the current scenario, we assumed that existing power plants are decommissioned at the end of their technical lifetime. For coal-fired power plants, a premature decommissioning pathway until 2040 is assumed and individually adjusted for each unit with regard to its technical parameters and national regulations, aligned as much as possible with the political guidelines on coal phase out known today. The portfolio of hydro power plants with storage or pump-storage capability is based on today's generation and announced construction or expansion projects under the assumption, that generators reaching the end of their lifetime are replaced with the same parameters.

In the presented case study, 555 existing or currently projected and 145 new built thermal power plants and 110 hydro power plants are operational in the scenario in 2030. The resulting thermal, hydro pump and hydro turbine capacity and the location of their connection to the transmission grid are shown in Fig. 1.

The case study year with input data for transmission grid as well as generation and demand is chosen as 2030 and the simulations are performed with an hourly time granularity. For the transmission grid part, a weekly optimization horizon of transmission grid without rolling horizon was chosen. For this, storage states of each unit at start and end time were fixed to the state given by the previously determined state after market-based dispatch.

[4]The results showed a mismatch between the calculated adequate capacity and the capacity in the scenario, which was significantly lower in some countries. The missing DSM modelling might explain some part of the mismatch.

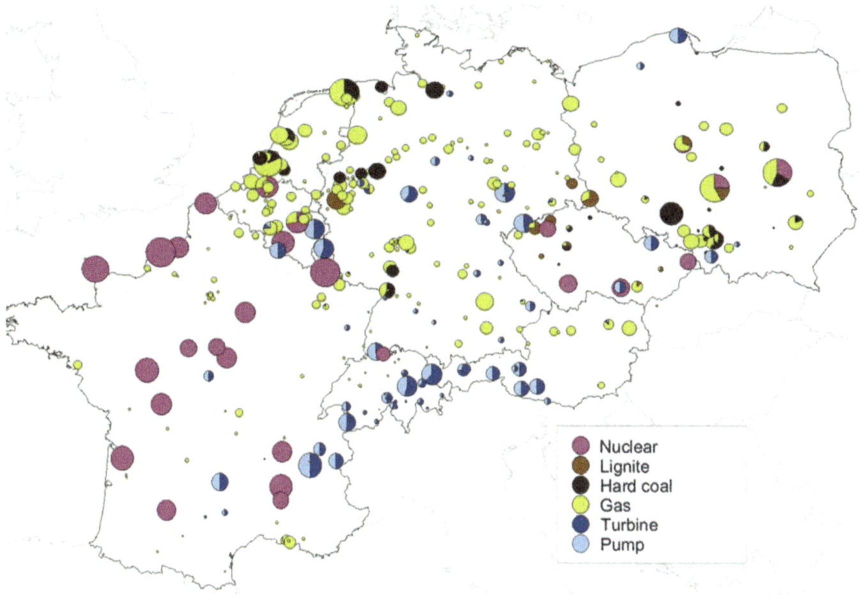

Fig. 1 Installed capacity per fuel type at the transmission grid substations. The circle volume is proportional of the legends circle volume of 2 GW

3.1 Transmission Grid Model

The transmission grid model contains the interconnected AC network of the same countries considered on the market side, including overhead lines and cables of voltage levels above 200 kV. For the studied area, this includes the nominal voltage levels of 220 and 380 kV. Additionally, high voltage DC (HVDC) lines are included in the dataset. HVDC lines connecting offshore wind generation are considered at the point of RES-E calculation and allocation and subsequently reduced from the dataset. All AC and HVDC lines are connected to busbars, which are assigned to georeferenced substations. The number of busbars at each substation is limited to the number of voltage levels present at the respective substation. In addition to the transformation between the extra high voltage levels, the transformation to the high voltage levels between 60 and 150 kV is considered and used for the connection of small-scale generation and demand as described in [18].

The data include the present state of the transmission grid as well as projected expansion measures in terms of deconstruction, replacement and construction of substations, busbars, lines and transformers until the year 2030. The technical data for each grid element was derived from publicly available sources whenever possible and otherwise approximated based on available information from predominant comparable equipment in the same geographical and organisational area. Among the sources used for completing the dataset are the static grid models of the transmission grid operators of the Central Western Europe Region (CWE), as well as the grid dataset of the TYNDP 2016 and Open Street Map (OSM). The future expansion of the national grid networks was obtained from various sources such as national network development plans of the respective grid operators and the projects listed in the TYNDP 2016. The model represents a snapshot of the projected grid topology in 2030 and includes a total of 3010 busbars, distributed over 1513 substations. The busbars are connected by 4103 AC lines and transformers, as well as six HVDC lines, with the exception of the interconnector between Belgium and Germany being located in the German market area. Furthermore, 237 reactive power compensation elements are located in the entire system, with either positive or negative reactive power provision. The snapshot of the transmission grid topology for 2030 is shown in Fig. 2.

The interconnection capacities between market areas are determined based on the available thermal capacities of interconnecting lines from the transmission grid dataset, assuming that 40% of the total thermal capacity of interconnecting AC lines between the market areas and the full thermal capacity of interconnection HVDC lines is being made available for market operation. Generators from non-intermittent sources are assumed to have reactive power provision capabilities of $\cos\varphi = 0.925$ w.r.t the installed generator capacity. Furthermore, RES-E with a nodal distribution of installed capacity as shown in Fig. 3 is assumed to be able to

Fig. 2 Map of the transmission 2030 used in the case study. Red denotes 380 kV lines, green 220 kV lines and purple HVDC lines

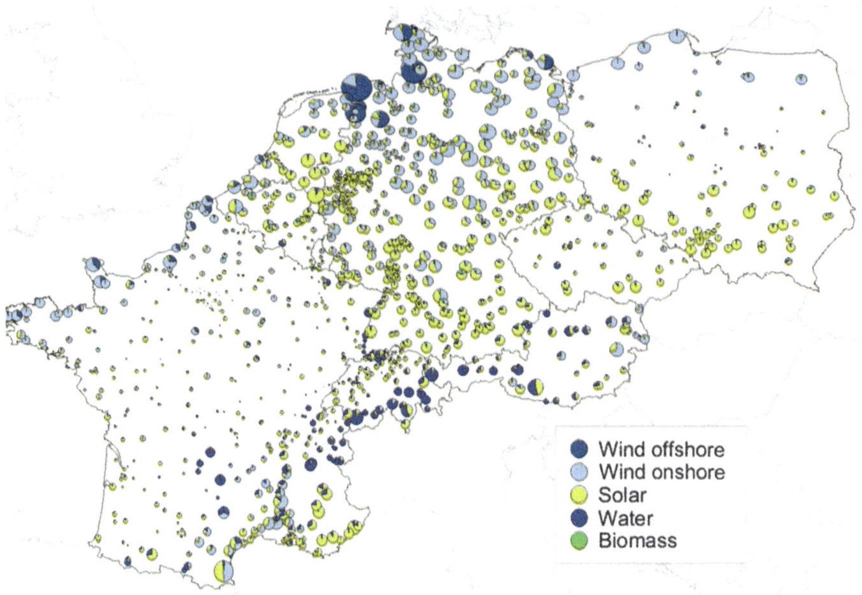

Fig. 3 Installed renewable capacity per type at the transmission grid substations. The circle volume is proportional to the legends circle volume of 2 GW

provide reactive power up to a level of $\cos\varphi = 0.95$ for all sources of renewable energy. With regard to the reactive power consumption, the reactive power demand in every market area and of every demand type is assumed to remain constant over time with a lagging power factor of $\cos\varphi = 0.95$. The maximum voltage deviation from nominal voltage is restricted to $\pm 10\%$ and grid components can be utilised up to their continuous thermal current restriction.

3.2 *Parameters for Flexibility Modelling*

Considering the modelling of new consumers, their flexibility and their regionalisation the following assumptions were made:

3.2.1 Heat Pumps in Households

The modelling of electricity consumed due to heat demand in households from HPs follows the approach of the Munich City Utilities (SMW) for calculating the standard load profile of HPs within their grid, based on the yearly household load and the temperature. Assuming a full flexibility ($\alpha^{flex} = 1, \quad \alpha^{+} = 1, \quad \alpha^{-} = 0$) within a 3 h horizon ($[00:00 - 03:00), [03:00 - 06:00), \ldots.$), the regionalisation follows the same approach as that of the household demand, restricted to single and two family buildings.

3.2.2 Heat Pumps in District Heating

We apply the same approach for modelling the generation profile and its flexibility as it is the case for HH-HP. Due to the availability of a heating grid and the possibility to store larger amounts of energy, a full flexibility within a 24 h horizon $[00:00-24:00)$ is assumed. The regionalisation is in general based on the capacity of combined heat and power (CHP) units within a country. In Germany furthermore the data of the German District Heating Association (AGFW) for the actual district heating grids demand and location where combined with the CHP database for a more detailed analysis.

3.2.3 BEV PBEV

Profiles are based on [14, 20] for a direct charging at arrival and adjusted based on Table 1.

The regionalisation corresponds to that of the household demand in all European countries except of Germany, where the vehicle registration numbers at NUTS 3 level were used as a distribution key instead of household number and electricity demand. Concerning the flexibility, we assumed the potential to shift the full demand ($\alpha^{flex} = 1, \ \alpha^{+} = 1, \ \alpha^{-} = 0$) within a 12 h horizon either during the day or the night ($[06:00-18:00)$, $[18:00:-06:00)$.

3.2.4 PV-BESS

In the current analysis, decentral BESS installed with PV on household level are modelled as simple battery storages and differ only in their sizing and regionalization from classical battery storages. Based on a linear relation between household demand and self-optimised PV-BESS capacity, obtained from a mixed integer optimization of PV-BESS sizing of households [21] in northern Germany and

Table 1 BEV parameter set

	Charging at home [kW]	Charging at work [kW]	Range [km]	Mean distance [km/a]	Demand [kWh/100 km]
2020	3,5	0	200	13.073	22
2030	5	5	300	13.071	22
2040	7	7	400	13.364	22
2050	11	11	400	13.481	22

southern Germany, an upper bound for the regional distribution of PV-BESS is calculated. By adjusting the BESS capacity according to the actual installed rooftop PV smaller 15kWp in single and two family buildings, the key for the final PV-BESS distribution is computed.

3.2.5 RES-E Flexibilities

For the flexibility of hydro storage and biomass generation we assumed $\alpha^{flex} = 0.5,\ \alpha^{+} = inf,\ \alpha^{-} = 0$, meaning that half of the generation profile is assumed to be flexible and restricted to the maximum value of the total time series of the flexible part. Furthermore, the potential to shift the generation within 24 h was assumed in case of biomass and of 168 h in case of hydro storage.

4 Results

4.1 Scenarios

In the following, we will present the results obtained from the simulations performed with the input dataset described above. The results are presented for four scenarios, which differ due to the amount of utilisation of the flexibilities both on the market and the transmission grid side. In this, we divide the available flexibilities into traditional flexible generation from hydro storage and into distributed flexibilities (HP, BEV, BESS, RES-E flexibilities) In the **Base** scenario, distributed flexibilities remain static according to their inelastic profile and time-dependent generation is not available for adjustment during grid operation. Thus, only hydro flexibilities are being utilised in the market operation and are subsequently fixed during the second step of grid operation simulation. The **NoFlex** scenario extends the Base scenario by adding hydro flexibilities to the available measures while resolving transmission grid congestion, keeping distributed flexibilities inelastic. The **Flex** scenario represents a further flexibilisation, with distributed flexibilities included in the market operation as described in Sects. 2.2, 2.3, 2.4, and 2.5 Lastly, in the **DynFlex** scenario the available BESS capacities after the market dispatch are added to the available adjustments during grid operation, increasing the number of storage units considered in the grid operation by 896 nodal distributed elements.

4.2 The Impact of Distributed Flexibilities

The yearly aggregated results are shown in Fig. 4. Transmission lines which reach their continuous thermal current limit during the time horizon are colored

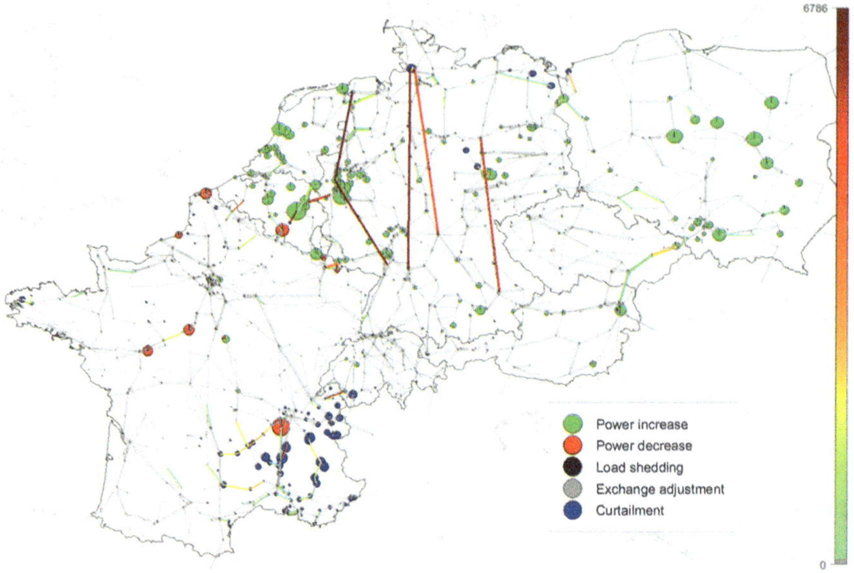

Fig. 4 Hours with maximum loading of transmission grid lines and average dispatch adjustment per type. The circle volume is proportional to the legends circle volume of 500 MW

according to the number of hours where this occurs, marking overloaded lines which require adjustments in the generation and demand dispatch in order to resolve the congestion. Congestions on the transformation level and voltage boundaries are omitted from Fig. 4. The highest number of fully loaded hours is found on the HVDC lines, implying that the start and end locations of these lines are generally well suited for relieving the stress on the AC grid. In the AC grid, areas with significant congestion can be found in Southern France, the border of France and Belgium, Northern Germany and the border area of Poland, the Czech Republic and Austria. Generally, the required decrease in generation at certain locations is reached by renewable curtailment rather than power decrease of thermal power plants. A trend of required power reduction in the southern part of the network can be observed over the entire year, while power increase is more focussed in the northern and eastern areas. When interpreting the results, possible discrepancies between the generation and demand scenario on the one hand and the grid model on the other hand have to be considered. For example, the grid expansion considered based on the TYNDP in France is significantly lower than the evolution of the grid in Germany in the time frame between today and 2030. Thus, the increase of both RES-E and new electricity demand types has an higher impact on a system that does not undergo a significant transformation.

The impact of the utilisation of distributed flexibilities on the security of supply can be seen in the reduction of the required load shedding in Table 2. While both the Base and the NoFlex scenario require load shedding in order to achieve a feasible

Table 2 Load shedding results

	Base	NoFlex	Flex	DynFlex
Hours with LS	240	136	11	9
Avg. hourly LS [MW]	20,48	9,00	0,78	0,78
Max. hourly LS [MW]	10043	6429,48	1794,14	2156,31

solution in more than 100 single hours during the year, this value is greatly decreased in the Flex scenario and even more in the DynFlex scenario. The main reason for this is the reduction in peak demand and generation due to the market-based dispatch of flexibilities as described in Sect. 2. Furthermore, the additional control capabilities – albeit with a small capacity over a longer time horizon – lead to less congestion events, which cannot be prevented without curtailing demand.

In the Base scenario, the average hourly positive dispatch adjustment requirement in the entire transmission grid area considered is 8.672 GW, with peaks reaching adjustments of up to 18.5 GW. As a permanent additional power provision is required in order to balance the transmission losses over lines and transformers, the positive adjustment does not decrease to zero even in cases without any violation of voltage or thermal limits. In the NoFlex scenario, the hourly requirement reduces by 48 MW while the peak increases to 20.1 GW. The hourly average decreases further to 8.54 GW in the Flex scenario and 8.497 GW in the DynFlex scenario, with the peak decreasing to 17.7 and 17.6 GW. The average negative dispatch adjustment required is not distorted by additional grid losses, over the scenarios it decreases by 3.6%, 2.1% and 1.0% from the NoFlex to the DynFlex scenario. Similarly to the positive adjustments, the negative yearly peak values are reaching the highest absolute values for the NoFlex scenario and the lowest for the DynFlex scenario.

Overall, the results show that as expected more flexibilities available both in the market and transmission grid lead to a lower amount of required redispatch. Yet, the different utilisations of both central and distributed flexibilities have a lower impact on the grid operation than expected. Among flexibilities the inclusion of large-scale hydro in the Flex scenario has the largest impact, as observed in the yearly values of negative dispatch adjustment. Operating distributed flexibilities according to central market signals leads to a small improvement in terms of transmission grid congestion. A similar observation can be made for the inclusion of BESS in the adjustable generation on the grid side. The comparably small difference between the scenarios might be explained by the discrepancies in the transmission grid model and the generation and demand data for 2030, with structural congestion accounting for a large part of the observed overloading events. However, the resulting security of supply is largely increased by utilising distributed flexibilities with load shedding mostly rendered unnecessary, as can be seen in Table 2. In order to improve the accuracy of findings for the total amount, more grid expansion in the input data or a better coordination of renewable expansion and demand increase on the one hand and the transmission grid on the other hand might be needed.

4.3 AC vs. DC Formulation

The comparison of hourly results of the DynFlex scenario with the results of an identical simulation performed with the DC restrictions in the load flow equations is shown in Fig. 5. For the sake of an equivalent comparison, grid losses are subtracted from the positive dispatch adjustment of the AC solution, and the demand of the DC model was increased by today's average national transmission grid losses for all market areas considered. Otherwise, both model formulations, restrictions and input data were kept consistent. The DC problem was solved using the commercial solver Gurobi. Even though the set of DC constraints form a relaxed formulation of the problem, the AC solution has a lower requirement for adjustments after correcting the AC results for the required adjustments to account for grid losses. This is due to the additional positive generation increase needed to account for grid losses in the AC case, which are part of the optimisation problem in the formulation chosen in this paper and thus contribute to relieving existing congestions. As a result they lower the additionally needed adjustment after the subtraction of the losses, leading to lower requirements, even though the linearised DC formulation is commonly referred to as a relaxation of the AC formulation. In the section up to 5 GW, the effect of voltage limits can be seen, which leads to higher AC adjustments as voltage is constant in the DC formulation. Overall, individual deviation in single hours can be significant while the entire trend and structure are similar. The correlation between the positive adjustment value series is 91.1%. This shows that the DC approach is suitable as an estimator for the AC equations for the investigation of general trends, while individual results and peaks might differ significantly due to the simplifications undertaken. For these cases, the AC formulation leads to better results.

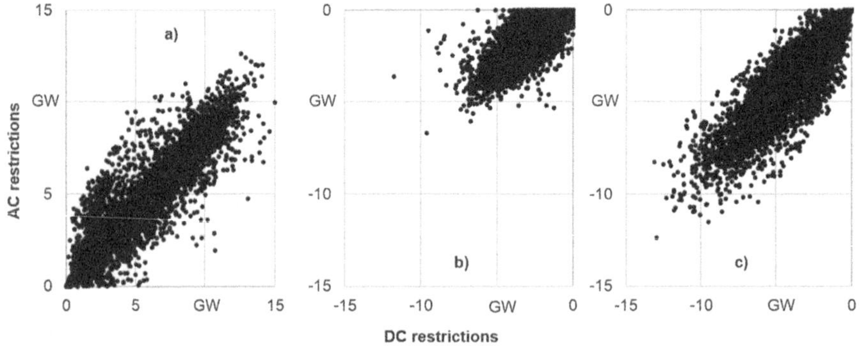

Fig. 5 (**a**) Distribution of positive adjustments for AC/DC results. (**b**) Distribution of negative adjustments of non-intermittent generators. (**c**) Distribution of negative adjustments of intermittent generators

5 Conclusion

In this paper, we presented a two-step approach to investigate the influence of distributed flexibilities on the operation of the transmission grid in the central European area. For this, we have developed a market optimisation formulation, which enables us to obtain dispatch results which include different kind of distributed flexible generation and demand sources. Subsequently, we analysed the effect of four scenarios with different implementation rates of flexibilities in market and grid operation based on a case study using data for the year 2030. Furthermore, we compared the results of our multi-period AC formulation to determine grid congestion with a linearised DC formulation.

We conclude based on the required adjustments over the simulation horizon of 8760 h, that a scenario like the DG scenario of the TYNDP, with high RES-E increase and the additional introduction of demand-side flexibilities into the electricity system is generally feasible for the anticipated transmission grid of the year 2030. Locally concentrated, larger adjustments can be explained by the gap between known grid expansion in selected countries and the ambitious targets of the chosen scenario. Also, distributed flexibilities show to have a strong effect on solving grid congestions that lead to load shedding, which is necessary in the Base scenario of our study in more than 200 h in order to achieve a feasible solution. This is almost entirely resolved when optimising distributed flexibilities with the market and enabling the adjustment of BESS during grid operation.

Furthermore, the results of the four scenarios investigated show that the impact of increased participation of both central and distributed flexibilities in the market and grid operation has a positive but comparably low influence on the overall required adjustment to operate the grid within its technical limits in our chosen case study. The largest change on the average adjustments required occurs when adding central storage units to the set of generators and consumers available for adjustment in the grid simulation. The comparison of the AC and DC results showed that the required adjustment can be lower in the AC case, if the provision of grid losses is included into the AC formulation and generation increase is performed at locations which are suitable for lowering stress on the grid. Additionally, the individual results of each hour showed a generally well reproduced trend using the DC relaxation, however single hours can differentiate significantly in both cases.

In the current state a simplified approach of modelling the flexible operation of shiftable demands and renewables based on their profile was integrated in the market model. In the future we are planning to integrate the decentral flexibilities modelling into the grid model und validate it against a more detailed representation of individual units. As demand and supply uncertainty have a major impact on the utilisation of storages and load shifting potentials, we are furthermore planning to extend our modelling to a stochastic optimization. Finally, the extension of the current modelling approach to a unit commitment problem would allow us to fully evaluate the contribution of flexibilities for the future power system.

Acknowledgements Manuel Ruppert and Rafael Finck gratefully acknowledge funding by the German Federal Ministry of Education and Research (BMBF) within the Kopernikus Project ENSURE "New ENergy grid StructURes for the German Energiewende".

Viktor Slednev kindly acknowledges the support for this work from the German Research Foundation (DFG) under the Project Number LE1432/14-2.

References

1. D. Keles, P. Jochem, R. McKenna, M. Ruppert, W. Fichtner, Meeting the modeling needs of future energy systems. Energy Technol. (2017). https://doi.org/10.1002/ente.201600607
2. X. Luo, J. Wang, M. Dooner, J. Clarke, Overview of current development in electrical energy storage technologies and the application potential in power system operation. Appl. Energy (2015). https://doi.org/10.1016/j.apenergy.2014.09.081
3. M. Beaudin, H. Zareipour, A. Schellenberglabe, W. Rosehart, Energy storage for mitigating the variability of renewable electricity sources: an updated review. Energy Sustain. Dev. (2010). https://doi.org/10.1016/j.esd.2010.09.007
4. A. Bloess, W.-P. Schill, A. Zerrahn, Power-to-heat for renewable energy integration: a review of technologies, modeling approaches, and flexibility potentials. Appl. Energy **212**, 1611–1626 (2018)
5. J. Hoppmann, J. Volland, T. Schmidt, V. Hoffmann, The economic viability of battery storage for residential solar photovoltaic systems – a review and a simulation model. Renew. Sustain. Energy Rev. (2014). https://doi.org/10.1016/j.rser.2014.07.068
6. D. Fischer, H. Madani, On heat pumps in smart grids: a review. Renew. Sustain. Energy Rev. (2017). https://doi.org/10.1016/j.rser.2016.11.182
7. D. Richardson, Electric vehicles and the electric grid: a review of modeling approaches, impacts, and renewable energy integration. Renew. Sustain. Energy Rev. (2013). https://doi.org/10.1016/j.rser.2012.11.042
8. D.P. Chassin, K. Schneider, C. Gerkensmeyer, GridLAB-D: an open-source power systems modeling and simulation environment, in *Exposition*, Chicago, pp. 1–5. https://doi.org/10.1109/TDC.2008.4517260
9. H.-K. Ringkjøb, P. Haugan, I. Solbrekke, A review of modelling tools for energy and electricity systems with large shares of variable renewables. Renew. Sust. Energ. Rev. (2018). https://doi.org/10.1016/j.rser.2018.08.002
10. C. de Jonghe, B.F. Hobbs, R. Belmans, Optimal generation mix with short-term demand response and wind penetration. IEEE Trans. Power Syst. (2012). https://doi.org/10.1109/TPWRS.2011.2174257
11. L. Baringo, A.J. Conejo, Transmission and wind power investment. IEEE Trans. Power Syst. (2012). https://doi.org/10.1109/TPWRS.2011.2170441
12. V. Krishnan, J. Ho, B.F. Hobbs, A.L. Liu, J.D. McCalley, M. Shahidehpour, Q.P. Zheng, Co-optimization of electricity transmission and generation resources for planning and policy analysis: review of concepts and modeling approaches. Energy Syst. (2016). https://doi.org/10.1007/s12667-015-0158-4
13. M. Ruppert, V. Slednev, V. Bertsch, W. Fichtner, The impact of microeconomic decisions in electricity market modelling on load flows in transmission grids, in *2016 13th International Conference on the European Energy Market (EEM). 2016 13th International Conference on the European Energy Market (EEM)*, 2016
14. P. Jochem, T. Kaschub, A.-G. Paetz, W. Fichtner, Integrating electric vehicles into the German electricity grid – an interdisciplinary analysis. World Electric Veh. J. (2012). https://doi.org/10.3390/wevj5030763
15. D. Madzharov, E. Delarue, W. D'haeseleer, Integrating electric vehicles as flexible load in unit commitment modeling. Energy (2014). https://doi.org/10.1016/j.energy.2013.12.009

16. M. Ruppert, V. Slednev, A. Ardone, W. Fichtner, Dynamic optimal power flow with storage restrictions using augmented lagrangian algorithm, in *2018 Power Systems Computation Conference*, Dublin (IEEE, 2018). https://doi.org/10.23919/PSCC.2018.8442993
17. S. Frank, S. Rebennack, An introduction to optimal power flow. Theory, formulation, and examples. IIE Trans. (2016). https://doi.org/10.1080/0740817X.2016.1189626
18. V. Slednev, M. Ruppert, V. Bertsch, W. Fichtner, N. Meyer-Hübner, M. Suriyah, T. Leibfried, P. Gerstner, M. Schick, V. Heuveline, Regionalizing input data for generation and transmission expansion planning models, in *Advances in Energy System Optimization*, ed. by V. Bertsch, W. Fichtner, V. Heuveline, T. Leibfried. Trends in Mathematics (Springer International Publishing, Cham, 2017), pp. 205–217
19. R. Hartel, V. Slednev, U. Yilmaz, A. Ardone, D. Keles, W. Fichtner, A. Eser, M. Klobasa, M. Kuhnbach, P. Manz, J. Globisch, R. Elsland, M. Wietschel, Dekarbonisierung des Energiesystems durch verstärkten Einsatz erneuerbaren Stroms im Wärme-, Verkehrs-und Industriesektor bei gleichzeitigen Stilllegungen von Kraftwerken – Auswirkungen auf die Versorgungssicherheit in Süddeutschland (2018)
20. V. Slednev, V. Bertsch, M. Ruppert, W. Fichtner, Highly resolved optimal renewable allocation planning in power systems under consideration of dynamic grid topology. Comput. Oper. Res. (2018). https://doi.org/10.1016/j.cor.2017.12.008
21. T. Kaschub, P. Jochem, W. Fichtner, Solar energy storage in German households: profitability, load changes and flexibility. Energy Policy (2016). https://doi.org/10.1016/j.enpol.2016.09.017

Part III
Managing Demand Response

A Discussion of Mixed Integer Linear Programming Models of Thermostatic Loads in Demand Response

Carlos Henggeler Antunes, Vahid Rasouli, Maria João Alves, Álvaro Gomes, José J. Costa, and Adélio Gaspar

Abstract In smart grids, it is expected that electricity retailers will offer time-differentiated tariffs with significant price variations in short periods. Consumers are then encouraged to engage in demand response strategies by making the most of their flexibility in the operation of end-use appliances to minimize the electricity bill without compromising on comfort. Air conditioning systems are particularly suited to be controlled, by profiting from the thermal inertia of building units. This paper presents novel mixed integer linear programming formulations to optimize the operation of a thermostatically-controlled air conditioning system, thoroughly discussing their main features and associated computational difficulties resulting from their combinatorial nature.

Keywords Mixed integer linear programming (MILP) · Thermostatic loads · Demand response

1 Introduction

Buildings in the commercial and residential sectors account for about 40% of the total energy consumption in European Union countries [1, 2]. Heating, Ventilation and Air Conditioning (HVAC) systems represent one of the most significant loads

C. H. Antunes (✉) · V. Rasouli · Á. Gomes
INESC Coimbra, Department of Electrical and Computer Engineering, University of Coimbra, Coimbra, Portugal
e-mail: ch@deec.uc.pt; vahid.rasouli@deec.uc.pt; agomes@deec.uc.pt

M. J. Alves
CeBER, Faculty of Economics and INESC Coimbra, University of Coimbra, Coimbra, Portugal
e-mail: mjalves@fe.uc.pt

J. J. Costa · A. Gaspar
ADAI-LAETA, Department of Mechanical Engineering, University of Coimbra, Coimbra, Portugal
e-mail: jose.costa@dem.uc.pt; adelio.gaspar@dem.uc.pt

V. Bertsch et al. (eds.), *Advances in Energy System Optimization*, Trends in Mathematics, https://doi.org/10.1007/978-3-030-32157-4_7

contributing to electrical energy consumption. HVAC systems are particularly suited to be controlled by making the most of: (i) the thermal inertia of spaces to be conditioned (leading to some time dissociation between the energy consumption and the energy service provided), and (ii) the flexibility of consumers' preferences regarding the provision of energy services in face of dynamic time-of-use tariffs (displaying some willingness to bear the indoor temperature somewhat farther from the reference settings for a limited period of time). Since time-differentiated tariff schemes are expected to become relevant commercial offers in electricity retail markets in smart grids, demand response actions should be developed to achieve optimal operation of HVAC systems. Optimal HVAC control can be beneficial for consumers (by reducing the energy bill without jeopardizing comfort), retailers (by managing buying and selling prices) and grid operators (by contributing to release the strain in distribution networks and enhancing the utilization of renewable sources). Consumer engagement in demand response programs can be made operational by means of automated energy management systems endowed with optimization algorithms and the capability to control appliances. Therefore, adequate and tractable models for optimal thermostat programming should be developed as operational tools (to be embedded with sensors and control) for demand response programs.

The potential of thermostatic loads for demand response actions has been exploited in the literature, due to their significant consumption and the possibility of being controlled. For this purpose, mathematical models have been proposed, in particular mixed integer linear programming (MILP) formulations, i.e. with integer and continuous variables. The authors in [3] consider heat pumps for heating residential buildings with a floor heating system using a linear state space model in an Economic Model Predictive Control framework. Thermal models are developed in [4] for a smart house for determining the value of thermal inertia in demand response dynamic pricing using a MILP formulation. The authors in [5] present a user-centric demand response control for scheduling the electric space heating load under price and load uncertainty to minimize a weighted-sum of the expected payment, loss of comfort, and financial risk of a customer while considering the end-user's preferences. The household thermal behavior is modeled by means of a two-capacity building model. An HVAC system is considered in [6] that is controlled (in combination with other type of loads, PV generation and storage) by a home energy management system, thus enabling residential consumers to participate in demand response programs. A price-based demand response strategy for multi-zone office buildings to optimize the energy cost of HVAC units and the thermal discomfort of occupants is formulated in [7] as a MILP model. The authors in [8] develop an approach based on a partial-differential equation model of thermal diffusion to determine the thermostat settings to minimize the electricity bill for a consumer with energy time-of-use and power prices, in which the optimal thermostat programming for HVAC is formulated as a constrained dynamic optimization problem. The authors in [9] consider the optimization of the investment and operation planning of a decentralized energy system, which is subject to different sources of uncertainties, encompassing photovoltaic generators and load flexibility using heat pumps in combination with thermal energy storage units for space heating and domestic hot water, which is tackled by two-stage stochastic programming.

This paper presents novel comprehensive MILP models in which the physical characteristics of a thermostatically-controlled air conditioning system are considered. These models have the common aim of determining the optimal thermostat operation using different forms of thermostat control. It will be discussed that the combinatorial nature of these models makes it difficult to solve them by a state-of-the-art exact solver. This fact impairs embedding realistic mathematical models in automated building energy management systems when a fine-grain time discretization of the planning period is considered.

The study is made in the perspective of the building owner or tenant, i.e. the one who aims to minimize the energy bill. These models can also be useful to optimize demand response strategies that can be of interest for the grid operator (possibly through the mediation of an aggregator of end-users' flexibility that uses this asset to participate in ancillary service provision markets). Thus, this paper adopts an Operational Research methodology focusing on the development of mathematical programming models.

The paper begins by developing a simple thermodynamic model to derive a time-discretized equation expressing the instantaneous indoor temperature (at instant t_i) as a function of the indoor temperature, the external temperature and the air conditioning system operation at the preceding instant (t_{i-1}). MILP models are then presented accounting for the hysteresis behavior of the thermostat between minimum and maximum temperatures defining the dead-band around a reference temperature (set-point). It is shown that, if just modelling this behavior, the model is solved in an acceptable computation time. However, if a minimum ON or OFF period is considered (also to avoid excessive switching), offering increased operation flexibility, the computation effort becomes impracticable.

In Sect. 2 the building thermal model is developed. MILP models of a thermostatically-controlled air conditioning system with different features are described in Sect. 3. Section 4 reports extensive computational experiments of the different MILP models. Conclusions are drawn in Sect. 5.

2 Development of the Building Thermal Model

A space, a building unit, is considered for being heated/cooled by means of an air conditioning (AC) system. Assuming the building unit as a (homogeneous) control volume, whose instantaneous indoor temperature is uniform and denoted by $\theta^{in}(t)$, the overall thermal energy balance on a heat rate basis (at some generic instant t) is [10]:

$$q^{ext}(t) + q^{g}(t) + q^{AC}(t) = C\frac{d\theta^{in}(t)}{dt} \tag{1}$$

for AC in heating mode, and

$$q^{ext}(t) + q^g(t) - q^{AC}(t) = C\frac{d\theta^{in}(t)}{dt} \tag{2}$$

for AC in cooling mode.

In these expressions: $q^{ext}(t)$ is the instantaneous external load [kW]; $q^g(t)$ is the instantaneous internal load (rate of heat generation, by occupants, lighting, appliances) [kW]; $C = \rho c_p V$ is the overall thermal capacity [kJ/°C], in which ρ and c_p are, respectively, the mass density [kg/m^3] and the specific heat at constant pressure [kJ/(kg.°C)] of indoor air, if no other thermal inertia of the building is considered, or some weighted values for indoors, and V is the volume of the building unit [m^3]; $q^{AC}(t)$ is the instantaneous air conditioning heating/cooling load [kW].

Considering a finite-difference approach over a time-step Δt to integrate equation (1) in a fully explicit time discretization – all variables at the "previous" instant t_{i-1}, except for temperature $\theta^{in}(t_i)$ at the "present" instant of Δt – it can be assumed that the rate of the energy storage, i.e. the right-hand side in (1) is given by $C\frac{\theta^{in}(t_i)-\theta^{in}(t_{i-1})}{\Delta t}$, and the external load $q^{ext}(t_i) \approx U.A\,[\theta^{ext}(t_{i-1}) - \theta^{in}(t_{i-1})]$. U is the (weighted-average) overall heat transfer coefficient of the building unit envelope [kW/(m^2·°C)] and A is the surface area of the envelope [m^2]. Therefore, $U.A$ represents the overall thermal conductance of the unit envelope [kW/°C].

Introducing the expressions above in (1), equations (3), (4), (5) are obtained:

$$U.A\left[\theta^{ext}_{t_{i-1}} - \theta^{in}_{t_{i-1}}\right] + q^g_{t_{i-1}} + q^{AC}_{t_{i-1}} = \frac{C}{\Delta t}(\theta^{in}_{t_i} - \theta^{in}_{t_{i-1}}) \tag{3}$$

$$\theta^{in}_{t_i} = \theta^{in}_{t_{i-1}} + \frac{U.A}{C}\Delta t\,(\theta^{ext}_{t_{i-1}} - \theta^{in}_{t_{i-1}}) + \frac{\Delta t}{C}\,(q^g_{t_{i-1}} + q^{AC}_{t_{i-1}}) \tag{4}$$

$$\theta^{in}_{t_i} = (1 - \frac{U.A}{C}\Delta t)\,\theta^{in}_{t_{i-1}} + (\frac{U.A}{C}\,\Delta t)\,\theta^{ext}_{t_{i-1}} + \frac{\Delta t}{C}\,(q^g_{t_{i-1}} + q^{AC}_{t_{i-1}}) \tag{5}$$

In a dynamic simulation of buildings, the internal thermal load q^g [kW] is usually defined according to the operation profile (e.g., daily/weekly profile of occupation, of lighting, etc.), i.e. all internal loads (thermal power) associated with the type of utilization of the building.

Without loss of generality for the application of the model, and for the sake of simplification in this context, the following assumptions are considered:

1. the internal heat loads are neglected: $q^g_{t_{i-1}} \approx 0$.
2. $q^{AC}_{t_{i-1}} = COP\ P^{demand}_{t_{i-1}} = COP\ P^{AC}_{nom} s_{t_{i-1}}$

where COP is the AC coefficient of performance, $P^{demand}_{t_{i-1}}$ and P^{AC}_{nom} are the power demand and the AC nominal power, respectively, and $s_{t_{i-1}}$ is the AC control variable

Table 1 Typical values of parameters α, β and γ for different discretization intervals of the planning period

Parameter	$\Delta t = 1\,\text{h}$	$\Delta t = 15\,\text{min}$	$\Delta t = 5\,\text{min}$	$\Delta t = 1\,\text{min}$
α	0.4276	0.8569	0.9523	0.99046
β	0.5724	0.1431	0.0477	0.00954
γ [°C/kW]	11.1111	2.7778	0.9259	0.1852

ON/OFF at instant t_{i-1}. Equation (5) can be rewritten as:

$$\theta_{t_i}^{in} = (1 - \frac{U.A}{C}\Delta t)\ \theta_{t_{i-1}}^{in} + (\frac{U.A}{C}\Delta t)\ \theta_{t_{i-1}}^{ext} + \frac{\Delta t}{C} COP\ P_{nom}^{AC} s_{t_{i-1}} \tag{6}$$

$$\theta_{t_i}^{in} = \alpha\ \theta_{t_{i-1}}^{in} + \beta\ \theta_{t_{i-1}}^{ext} + \gamma\ P_{nom}^{AC} s_{t_{i-1}} \tag{7}$$

$$\alpha = 1 - \beta \tag{8}$$

$$\beta = \frac{U.A}{C}\Delta t \tag{9}$$

$$\gamma = \frac{\Delta t}{C} COP \tag{10}$$

Therefore, since COP, P_{nom}^{AC}, Δt and C are always positive quantities, the sign in equation (10) is positive as this equation has been derived from (1) for heating mode. Otherwise, $\gamma = -\frac{\Delta t}{C} COP$ (negative) for cooling mode.

The coefficients α, β and γ derive from the building characteristics (area, envelope, etc.) and the COP of the AC. Considering a building unit with a 225 m^2 floor area and 3 m height, and the properties of the air inside (density of 1.19 kg/m^3 and specific heat of 1.007 kJ/(kg.°C)), the air thermal capacity is $C \approx 810$ kJ/°C. Taking U-values of 0.35 and 0.30 W/(m^2.°C) for the building facades and roof, respectively, a value of $U.A \approx 0.129$ kW/°C is obtained. Assuming a $COP = 2.5$, typical values of parameters α, β and γ are displayed in Table 1 for different discretization intervals of the planning period.

3 Mathematical Models of a Thermostatically-Controlled AC System

This section presents different MILP models aimed at determining the optimal operation of the AC within a planning period to minimize energy costs, considering distinct forms of control. Although other types of loads could be considered (such as shiftable and interruptible appliances), the goal herein is to focus on the mathematical modelling of the AC operation. Shiftable loads include dishwashers, laundry machines and clothes driers, i.e. appliances for which the operation cycle cannot be interrupted once initiated. The supply of interruptible loads, which include electric water heaters and the battery of electric vehicles, can be interrupted within preferred time slots provided a certain amount of energy is supplied. In general, shiftable and interruptible loads require MILP models, which can be solved very fast by state-of-the-art solvers (e.g. Cplex), although requiring many binary variables. Examples of such models are the ones presented in [11] and [12]. However, the MILP modelling of the thermostat behavior is more complex and imposes a higher computational effort to obtain the optimal solution. In particular, MILP models are presented accounting for the hysteresis behavior of the thermostat between minimum and maximum temperatures defining the dead-band around a reference temperature (set-point) established by the user according to his comfort preferences.

The planning period consists of T time intervals $t = 1, \ldots, T$ of a given duration; this discretization can be, for instance, 15-, 5- or 1-min. The length of the time interval is denoted by Δt, in hours. For instance, if a 1-min discretization is used then $\Delta t = 1/60$ h.

Figure 1 displays the behavior of the thermostat of an AC in heating mode. The hysteresis operation of the thermostat prevents excessive switching when the indoor temperature is around the set-point, which may be established as the mid-point within the dead-band defined by the comfort range of indoor temperature $[\theta^{min},\theta^{max}]$ specified by the user. These comfort bounds may depend on t, i.e. may be different during the planning period ($[\theta_t^{min}, \theta_t^{max}], t = 1, \ldots, T$); for the sake of clarity, we will consider the comfort bounds constant throughout the planning period in the formulations presented below. Hysteresis determines the AC control variable:

- $s_t = 1$, i.e. the AC continues ON as the indoor temperature θ_t^{in} increases above θ^{min} until it reaches θ^{max} (a in Fig. 1);
- $s_t = 0$, i.e. the AC continues OFF as the indoor temperature decreases below θ^{max} until it reaches θ^{min} (b in Fig. 1).

3.1 *Modelling the Thermostat Behavior*

Model 1A forces $s_t = 1$ when the indoor temperature (θ_t^{in}):

- is below the lower bound of the comfort band ($\theta_t^{in} < \theta^{min}$) *OR*

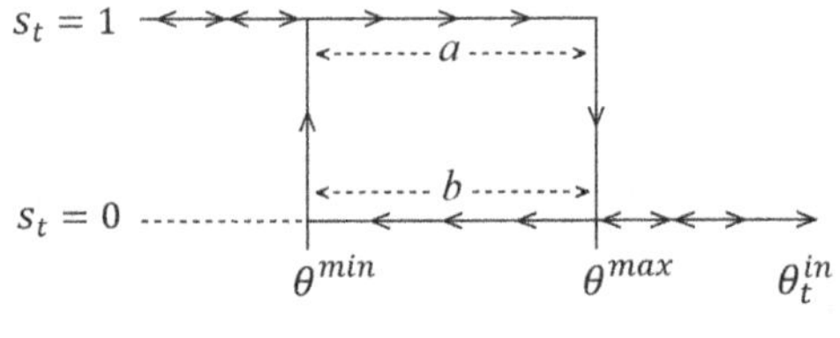

Fig. 1 Behavior of an AC thermostat for heating mode

- is within the comfort band ($\theta^{min} \leq \theta_t^{in} \leq \theta^{max}$) *AND* in the previous instant the AC was also ON ($s_{t-1} = 1$)

Model 1A (heating mode):

$$min \sum_{t=1}^{T} [\, c_t \, P_{nom}^{AC} \, s_t] \Delta t \tag{11}$$

s.t.:

$$\theta_t^{in} = \alpha\, \theta_{t-1}^{in} + \beta\, \theta_{t-1}^{ext} + \gamma\, P_{nom}^{AC}\, s_{t-1} \qquad t = 1, \ldots, T \tag{12}$$

$$\theta_t^{in} \geq \theta^{max} - M y_t \qquad t = 1, \ldots, T \tag{13}$$

$$\theta_t^{in} \geq \theta^{min} - M w_t \qquad t = 1, \ldots, T \tag{14}$$

$$s_{t-1} \leq w_t \qquad t = 1, \ldots, T \tag{15}$$

$$y_t + w_t - s_t \leq 1 \qquad t = 1, \ldots, T \tag{16}$$

$$s_t \in \{0, 1\}, \, y_t \in \{0, 1\}, \, w_t \in \{0, 1\}, \quad t = 1 \ldots, T \tag{17}$$

$$\theta_t^{in} \geq 0 \qquad t = 1, \ldots, T \tag{18}$$

The binary variables s_t control the thermostat switching:

$$s_t = \begin{cases} 0 & \text{if the AC is OFF} \\ 1 & \text{if the AC is ON} \end{cases}$$

The auxiliary binary variables y_t and w_t establish the consistency conditions associated with thermostat switching to the ON position ($s_t = 1$):

$$y_t = \begin{cases} 1 & \text{if } \theta_t^{in} < \theta^{max} \\ 0 & \text{otherwise} \end{cases}$$

$$w_t = \begin{cases} 1 & \text{if } \theta_t^{in} < \theta^{min} \; OR \; s_{t-1} = 1 \\ 0 & \text{otherwise} \end{cases}$$

M is a positive large number, which is used in (13) and (14) to enforce the logical conditions. M must be larger than any temperature value; for instance, $M = 100$

(or any higher value) is suitable as temperatures are in °C. The constraints of Model 1A ensure that:

1. if $\theta_t^{in} < \theta^{min}$, then $y_t = 1$ and $w_t = 1$ by (13) and (14); hence, by $y_t + w_t - s_t \leq 1$ (16), $s_t = 1$;
2. if $\theta^{min} \leq \theta_t^{in} < \theta^{max}$ and $s_{t-1} = 1$, then $y_t = 1$ and $w_t = 1$ by (13) and (15); hence, by $y_t + w_t - s_t \leq 1$ (16), $s_t = 1$;
3. in other cases, s_t is free to be 0 or 1. The objective function pushes these variables to 0, in order to turn the AC to the OFF state and minimize cost.

Additional input information to be specified include:

- The temperatures defining the thermostat dead-band, i.e. the minimum temperature θ^{min} (°C) below which the thermostat should be ON ($s_t = 1$) and the maximum temperature θ^{max} (°C) above which the thermostat should be OFF ($s_t = 0$).
- The initial indoor temperature θ_0^{in} (°C) and the initial status (0, 1) of the thermostat s_0.
- A forecast of external temperatures θ_t^{ext} (°C).
- The nominal power of the AC, P_{nom}^{AC} (kW).
- The cost of energy c_t (€/kWh) for $t = 1, \ldots, T$, in which the coefficients c_t derive from a time-of-use tariff (a commercial offer of a retailer selected by the user).

This model assumes that the "natural state" of the control variable s_t is 0 (since it has a positive coefficient in the objective function to be minimized). The objective function forces the binary variables to 0 whenever the constraints do not impose these variables to be 1. However, in this model it may happen that, when indoor temperature drops below the upper bound temperature (θ^{max}) after having been above it (and hence $s_t = 0$), the control variable switches from 0 to 1 within the thermostat comfort band (note that there are no constraints forcing $s_t = 0$ because this would be the normal state of the variable). This may be beneficial for the objective function by avoiding switching on at some later time intervals when the electricity cost is higher. Therefore, Model 1A does not replicate exactly the thermostat hysteresis behavior. This model guarantees the maintenance of the ON state within the comfort dead-band (*a* in Fig. 1) but not the maintenance of the OFF state (*b* in Fig. 1). This freedom to switch ON to benefit the cost objective function is the reason why this model takes so much computation time. A model forcing the control variables to 0 or 1 according to the thermal balance equation and thermostat hysteresis, i.e. a rule-based system modelling the thermostat switching, is almost instantaneously solved. This is implemented in Model 1B. In addition to Model 1A, Model 1B (also for heating mode) explicitly forces $s_t = 0$ when the indoor temperature (θ_t^{in}):

- is above the upper bound of the comfort band ($\theta_t^{in} > \theta^{max}$) *OR*
- is within the comfort band ($\theta^{min} \leq \theta_t^{in} \leq \theta^{max}$) *AND* in the previous instant the AC was also OFF ($s_{t-1} = 0$) (*b* in Fig. 1)

Model 1B (heating mode):

$$min \sum_{t=1}^{T} [\, c_t \, P_{nom}^{AC} \, s_t] \Delta t \tag{19}$$

s.t.:

$$\theta_t^{in} = \alpha\, \theta_{t-1}^{in} + \beta\, \theta_{t-1}^{ext} + \gamma\, P_{nom}^{AC}\, s_{t-1} \qquad t = 1, \ldots, T \tag{20}$$

$$\theta_t^{in} \geq \theta^{min} - M s_t \qquad t = 1, \ldots, T \tag{21}$$

$$\theta_t^{in} \leq \theta^{min} + M z_t \qquad t = 1, \ldots, T \tag{22}$$

$$\theta_t^{in} \geq \theta^{max} - M y_t \qquad t = 1, \ldots, T \tag{23}$$

$$\theta_t^{in} \leq \theta^{max} + M(1 - s_t) \qquad t = 1, \ldots, T \tag{24}$$

$$z_t + y_t - s_{t-1} + s_t \leq 2 \qquad t = 1, \ldots, T \tag{25}$$

$$z_t + y_t + s_{t-1} - s_t \leq 2 \qquad t = 1, \ldots, T \tag{26}$$

$$s_t \in \{0, 1\}, y_t \in \{0, 1\}, z_t \in \{0, 1\} \quad t = 1, \ldots, T \tag{27}$$

$$\theta_t^{in} \geq 0 \qquad t = 1, \ldots, T \tag{28}$$

The binary variables s_t keep the same meaning as in Model 1A, i.e. they control the thermostat switching.

The auxiliary binary variables y_t and z_t establish the consistency conditions associated with thermostat switching to the ON ($s_t = 1$) and OFF ($s_t = 0$) positions:

$$y_t = \begin{cases} 1 & \text{if } \theta_t^{in} < \theta^{max} \\ 0 & \text{otherwise} \end{cases}$$

$$z_t = \begin{cases} 1 & \text{if } \theta_t^{in} > \theta^{min} \\ 0 & \text{otherwise} \end{cases}$$

Constraints (21) impose that $s_t = 1$ if $\theta_t^{in} < \theta^{min}$. Constraints (22), (23) together with (25), (26) impose that s_t keeps the value of s_{t-1} within the comfort band. Constraints (24) impose that $s_t = 0$ if $\theta_t^{in} > \theta^{max}$.

However, this model is just the mathematical formulation of logical implications, which establish the values that the control variable should have according to the indoor temperature θ_t^{in} and thus no optimization is really at stake. That is,

s_t variables are rigidly determined by the conditions that establish θ_t^{in} and the thermostat operation (Fig. 1). In this case, there is a single feasible solution (the one that complies with the thermostat hysteresis switching rules), so obtaining the solution to this model is almost instantaneous.

3.2 Modelling Minimum ON/OFF Duration Periods

A new model (Model 2) has been developed to offer the possibility of controlling the AC to make the most of time-differentiated tariffs, i.e. being ON during lower electricity price periods when it is not strictly necessary to heat the building space so that thermal comfort is achieved even limiting the time ON when electricity prices are higher. In face of this freedom given to the model, it is necessary to inhibit excessive switching, which is accomplished by imposing a new set of constraints. In the comfort band, the AC control variables s_t are determined to minimize costs in face of dynamic tariffs, but once there is a switch from 1 to 0 or from 0 to 1, then the new state ON/OFF should be kept for at least d time intervals (these minimum ON/OFF periods could be different for each state). These new constraints establish:

- (33): if there is a change from $s_{t-1} = 1$ to $s_t = 0$, then the subsequent s_t should be equal to 0 during d time intervals.
- (34): if there is a change from $s_{t-1} = 0$ to $s_t = 1$, then the subsequent s_t should be equal to 1 during d time intervals.
- (35) and (36): in the last d time intervals of the planning period, if there is a switch from 1 to 0 or from 0 to 1, then this position must be kept until T.

This model is also quite demanding from the computation time point of view due to offering the possibility of switching from 1 to 0 or from 0 to 1 in the comfort band.

Model 2 (heating mode):

$$min \sum_{t=1}^{T} [\, c_t \, P_{nom}^{AC} \, s_t] \Delta t \tag{29}$$

s.t.:

$$\theta_t^{in} = \alpha \, \theta_{t-1}^{in} + \beta \, \theta_{t-1}^{ext} + \gamma \, P_{nom}^{AC} \, s_{t-1} \qquad t = 1, \ldots, T \tag{30}$$

$$\theta_t^{in} \geq \theta^{min} - M s_t \qquad t = 1, \ldots, T \tag{31}$$

$$\theta_t^{in} \leq \theta^{max} + M(1 - s_t) \qquad t = 1, \ldots, T \tag{32}$$

$$1 - s_t + s_{t-1} + \frac{1}{d} \sum_{j=t}^{t+d-1} s_j \leq 2 \quad t = 1, \ldots, T - d + 1 \tag{33}$$

$$1 - s_t + s_{t-1} + \frac{1}{d} \sum_{j=t}^{t+d-1} s_j \geq 1 \quad t = 1, \ldots, T - d + 1 \tag{34}$$

$$1 - s_t + s_{t-1} + \frac{1}{T - t + 1} \sum_{j=t}^{T} s_j \leq 2 \quad t = T - d + 2, \ldots, T \tag{35}$$

$$1 - s_t + s_{t-1} + \frac{1}{T - t + 1} \sum_{j=t}^{T} s_j \geq 1 \quad t = T - d + 2, \ldots, T \tag{36}$$

$$s_t \in \{0, 1\} \quad t = 1, \ldots, T \tag{37}$$

$$\theta_t^{in} \geq 0 \quad t = 1, \ldots, T \tag{38}$$

3.3 *Modelling an Inverter AC*

Inverter technology AC appliances have growing acceptance because of increased efficiency, extended life and elimination of abrupt load and temperature variations. In this type of appliances, a variable-frequency drive adjusts the compressor and the cooling/heating output. The previous models can be extended to accommodate inverter units, for instance considering they can be operated at different levels with respect to the nominal power. This can be modelled by introducing additional binary decision variables:

$$\delta_t^r = \begin{cases} 1 & \text{if the AC is operating at load level } r \\ & (r = 1, \ldots, R) \text{ at time } t \in T \\ 0 & \text{otherwise} \end{cases}$$

For instance, for $R = 5$ with the power levels 20–40–60–80–100% of the nominal power, the AC operation is modelled in Model 3. In this model, θ^{min} and θ^{max} should be interpreted as absolute bounds the user is willing to endure. Note that the sets of constraints in the previous models could also have been used here (for simplification purposes they were omitted). p_t^{AC} is the power required to operate the AC at time t of the planning period, i.e. the AC is either OFF ($p_t^{AC} = 0$) or supplied at one of the specified power levels.

Model 3 (heating mode):

$$min \sum_{t=1}^{T} [\, c_t \, P_t^{AC}]\Delta t \tag{39}$$

s.t.:

$$\theta_t^{in} = \alpha\, \theta_{t-1}^{in} + \beta\, \theta_{t-1}^{ext} + \gamma\, P_t^{AC} \qquad t = 1, \ldots, T \tag{40}$$

$$p_t^{AC} = (0.2\, \delta_t^1 + 0.4\, \delta_t^2 + 0.6\, \delta_t^3 + 0.8\, \delta_t^4 + \delta_t^5)\, P_{nom}^{AC} \qquad t = 1, \ldots, T \tag{41}$$

$$\delta_t^1 + \delta_t^2 + \delta_t^3 + \delta_t^4 + \delta_t^5 \le 1 \qquad t = 1, \ldots, T \tag{42}$$

$$\theta_t^{in} \ge \theta^{min} \qquad t = 1, \ldots, T \tag{43}$$

$$\theta_t^{in} \le \theta^{max} \qquad t = 1, \ldots, T \tag{44}$$

$$\delta_t^r \in \{0, 1\} \qquad r = 1, \ldots, 5; \quad t = 1, \ldots, T \tag{45}$$

$$p_t^{AC} \ge 0 \qquad t = 1, \ldots, T \tag{46}$$

3.4 Modelling the Cost vs. Comfort Trade-Off

If the energy price at instant t, c_t, is higher than a given threshold, c^g, and the user is willing to accept the indoor temperature some degrees, θ^g, below the minimum of the comfort band θ^{min} (or above the maximum θ^{max} in cooling mode), this can be modelled by defining the auxiliary variables y_t and the following constraints:

$$y_t = \begin{cases} 1 & \text{if } c_t \ge c^g \\ 0 & \text{otherwise} \end{cases}$$

$$\theta^{min} - y_t\, \theta^g \le \theta_t^{in} \le \theta^{max} \qquad t = 1, \ldots, T \tag{47}$$

$$c^g - c_t + M y_t \ge 0 \qquad t = 1, \ldots, T \tag{48}$$

$$c^g - c_t - M(1 - y_t) \le 0 \qquad t = 1, \ldots, T \tag{49}$$

The features of the models described above have been presented separately but they may be combined. For instance, the different power levels in Model 3 can be combined with the thermostat modelling of Models 1 and 2 and/or with the cost vs. comfort trade-off.

In the next section, illustrative results of the exact solution (whenever possible with a given computation budget) of these models using a MIP solver are presented drawing attention to the computational effort.

4 Illustrative Results

4.1 Data

A planning period of 24 h with a time discretization of 1 min is adopted, i.e. 1440 time steps, which imposes a heavier computational burden. The thermostat deadband is defined by $\theta^{min} = 20$°C and $\theta^{max} = 24$°C. The initial indoor temperature $\theta_0^{in} = 18$°C in all models, except in model 3 in which it is $\theta_0^{in} = 20$°C. The initial thermostat status $s_0 = 0$. The external temperatures θ_t^{ext}(°C) for a winter day (from a meteorological station in Coimbra, central Portugal, on January 1st, 2012) are available at http://www.uc.pt/en/org/inescc/publications/files/temp_out_winter.xlsx.

The nominal power of the AC, $P_{nom}^{AC} = 1.5$ kW. The time-differentiated energy prices c_t(€/kWh), considering 10 sub-periods, are given in Table 2. Experiments have been made on an Intel(R) Core(TM) i5, 3.33 GHz computer with GAMS ver. 24.0.2 and Cplex ver. 12.5.0.0.

4.2 Results

A computational budget of 7200 s or relative MIP gap of 0.5% was established. The behavior of thermostat switching as well as the evolution of indoor temperature are displayed in Figs. 2 and 3 for Model 1A and Model 1B, respectively. The cost of the corresponding optimal solutions for Model 1A and Model 1B, as well as the dimension of the models and the Cplex time to obtain these solutions are given in Table 3. A minimum duration ON/OFF after a thermostat switching occurs is considered with $d = 3$ and $d = 5$ min (Model 2). The behavior of thermostat switching and the evolution of indoor temperature are displayed in Figs. 4 and 5 for Model 2 with $d = 3$ and $d = 5$ min, respectively. The cost of the corresponding optimal solutions for $d = 3$ min and $d = 5$ min, as well as the dimension of the model and the Cplex time to obtain these solutions are given in Table 4. Considering that the AC can operate at power levels 20–40–60–80–100% of P_{nom}^{AC} in Model 3, the cost of the optimal solution as well as the dimension of the model and the Cplex time to obtain these solutions are given in Table 5. The corresponding solution of the AC operation and indoor temperature variation is displayed in Fig. 6.

Table 2 Electricity prices c_t(€/kWh) charged by the retailer, in each sub-period $P_i, i \in \{1, \ldots, 10\}$ (24-h planning period discretized in 1440 1-min intervals)

Price sub-period	P_1 [1,120]	P_2 [121,360]	P_3 [361,450]	P_4 [451,630]	P_5 [631,750]	P_6 [751,840]	P_7 [841,1020]	P_8 [1021,1140]	P_9 [1141,1260]	P_{10} [1261,1440]
c_t(€/kWh)	0.1	0.075	0.12	0.24	0.21	0.18	0.21	0.24	0.30	0.15

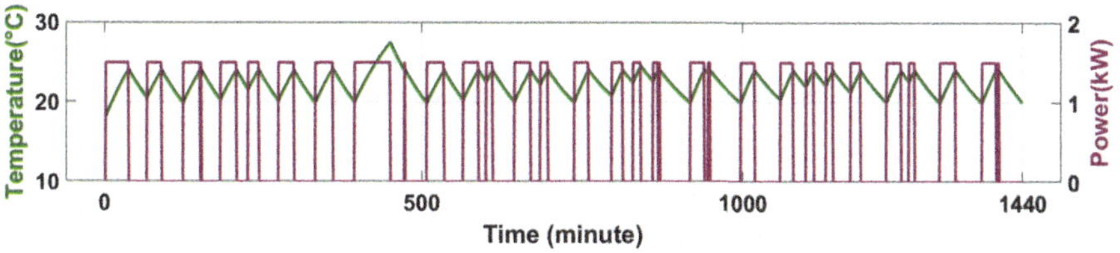

Fig. 2 Optimal AC operation and indoor temperature for Model 1A

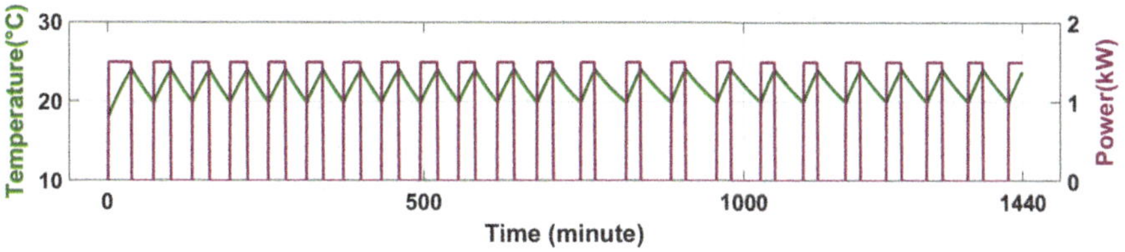

Fig. 3 Optimal AC operation and indoor temperature for Model 1B

Table 3 Optimal solutions, problem dimension and Cplex time (Model 1A and 1B)

Model name	Optimal solution (€)	Relative gap (%)	Cplex time (s)	N. binary variables	N. continuous variables	N. constraints
Model 1A	2.416125	43.47	7200	4320	1440	7199
Model 1B	2.465125	0	0.01	4320	1440	10078

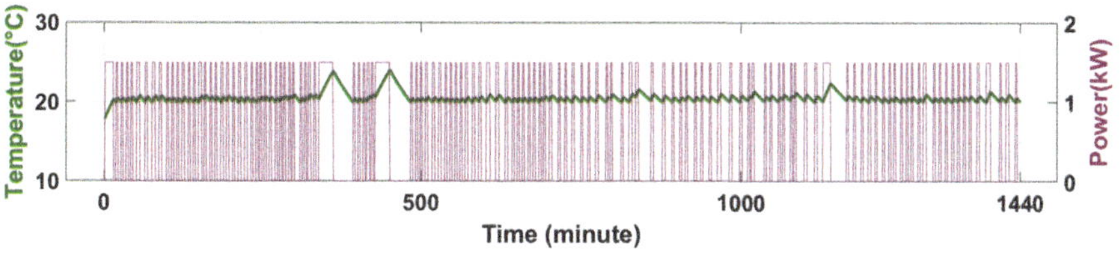

Fig. 4 Optimal AC operation and indoor temperature for Model 2 with $d = 3$ min

The frequency of switching in Model 1A is higher than in Model 1B due to the freedom given by that model to minimize cost. In Model 1A, the AC is ON for a longer time in pricing sub-period P_3 to account for the increase in price in subsequent sub-periods (Fig. 2). Note that the solution presented in Fig. 2 and Table 3 for Model 1A still presents a MIP gap of 43.47% after 2 h of computation, due to the combinatorial difficulties of this model. Model 1B, as expected, is solved to optimality instantly.

The temperature peaks in Figs. 4, 5 and 6 are due to the capability of the models to accommodate for comfort by making the most of lower price sub-periods. The MIP gaps in Model 2 (Table 4) are still significant after 2 h of computation.

The optimal solution to Model 3 is better than the solutions obtained for the other models due to the possibility of the inverter AC to work at different power levels.

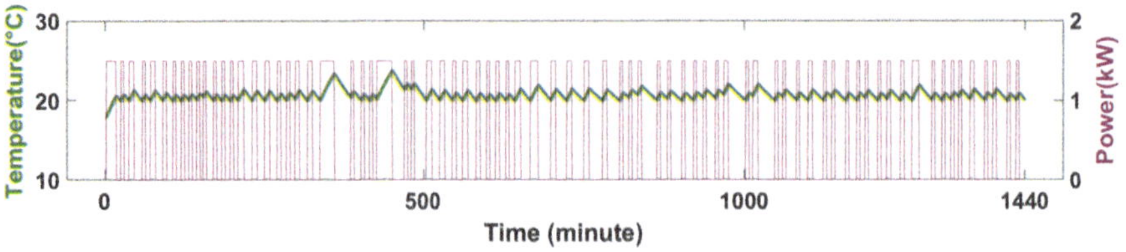

Fig. 5 Optimal AC operation and indoor temperature for Model 2 with $d = 5$ min

Table 4 Optimal solutions, problem dimension and Cplex time (Model 2)

Model name	Optimal solution (€)	Relative gap (%)	Cplex time (s)	N. binary variables	N. continuous variables	N. constraints
Model 2 ($d = 3$ min)	2.015875	6.95	7200	1440	4320	8636
Model 2 ($d = 5$ min)	2.07725	10.62	7200	1440	4320	8636

Table 5 Optimal solutions, problem dimension and Cplex time (Model 3)

Model name	Optimal solution (€)	Relative gap (%)	Cplex time (s)	N. binary variables	N. continuous variables	N. constraints
Model 3	1.94445	0.57	7200	7200	2880	7200

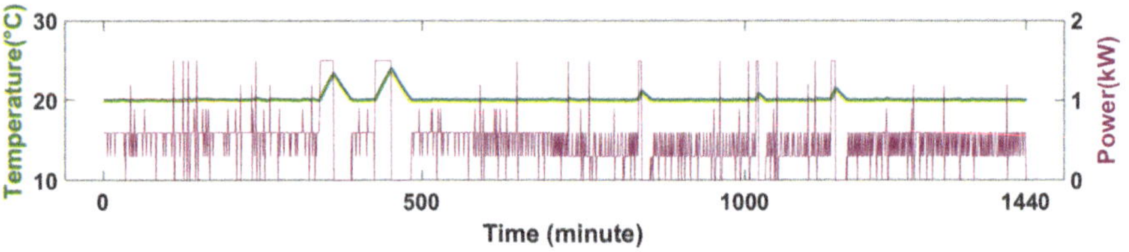

Fig. 6 Optimal AC operation and indoor temperature for Model 3

5 Conclusion

This paper presented a set of mixed integer linear programming models in order to optimize the operation of a thermostatically-controlled air conditioning system (AC), aiming at minimizing the energy costs. These models capture different physical control characteristics:

- Model 1A models the hysteresis operation of the thermostat (in heating mode), preventing from excessive switching when the indoor temperature is around a set-point. This model gives, however, some freedom to thermostat switching when the indoor temperature is within the dead-band, enabling to change from OFF to ON if this benefits the cost objective function.
- Model 1B removes the flexibility given by the previous model, forcing the AC to keep its state when the indoor temperature is within the dead-band. This model implements, in a MILP formulation, a rule-based system of logical implications.

The feasible region is reduced to one solution and the model can be solved very quickly, unlike Model 1A which is computationally very demanding.
- Model 2 gives some flexibility while inhibiting excessive switching, by considering that, when the indoor temperature is within the dead-band, the thermostat must keep its state (ON or OFF) for at least a predefined number of time intervals. This model is also quite computationally demanding.
- Model 3 introduces a new feature that can be used to extend the previous models: it models inverter units, considering that the AC can operate at different levels of nominal power.

Modelling cost reduction at the expense of accepting worsening the indoor temperature by some degrees was also introduced in this paper. It has been shown that the combinatorial nature of these models imposes a severe computation burden, which in some cases impairs obtaining the optimal solutions in a practical, acceptable computational time by a state-of-the-art solver. Therefore, this work lays the foundations for understanding those computational difficulties and gives insights for the development of other approaches, namely dedicated meta-heuristics customized for the features of the models, in order to offer good solutions with a suitable computational effort.

Acknowledgements This work was partially supported by projects UID/MULTI/00308/2013 and by the European Regional Development Fund through the COMPETE 2020 Programme, FCT – Portuguese Foundation for Science and Technology within projects ESGRIDS (POCI-01-0145-FEDER-016434) and MAnAGER (POCI-01-0145-FEDER-028040).

References

1. P. Bertoldi, F. Diluiso, L. Castellazzi, N. Labanca, T. Ribeiro Serrenho, *Energy Consumption and Energy Efficiency Trends in the EU-28 2000–2015*, EUR 29104 EN, Publications Office of the European Union, Luxembourg, 2018, JRC110326
2. D. D'Agostino, P. Paolo Zangheri, L. Castellazzi, Towards nearly zero energy buildings in Europe: a focus on retrofit in non-residential buildings. Energies **10**(1), 117 (2017)
3. R. Halvgaard, N.K. Poulsen, H. Madsen, J.B. Jørgensen, Economic model predictive control for building climate control in a smart grid, in *IEEE PES Innovative Smart Grid Technologies (ISGT) Conference*, Washington, DC, 2012
4. H. Wang, K. Meng, F. Luo, Z.Y. Dong, G. Verbič, Z. Xu, K.P. Wong, Demand response through smart home energy management using thermal inertia, in *Australasian Universities Power Engineering Conference (AUPEC)*, Hobart, 2013
5. M. Ali, J. Jokisalo, K. Siren, A. Safdarian, M. Lehtonen, A User-centric demand response framework for residential heating, ventilation, and air-conditioning load management. Electr. Power Compon. Syst. **44**(1), 99–109 (2016)
6. M. Shafie-Khah, P. Siano, A Stochastic home energy management system considering satisfaction cost and response fatigue. IEEE Trans. Ind. Inf. **14**(2), 629–638 (2018)
7. Y. Kim, Optimal price-based demand response of HVAC systems in multi-zone office buildings considering thermal preferences of individual occupants. IEEE Trans. Ind. Inf. **14**(11), 5060–5073 (2018)

8. R. Kamyar, M. Peet, Optimal thermostat programming for time-of-use and demand charges with thermal energy storage and optimal pricing for regulated utilities. IEEE Trans. Power Syst. **32**(4), 2714–2723 (2017)
9. H. Schwarz, V. Bertsch, W. Fichtner, Two-stage stochastic, large-scale optimization of a decentralized energy system – a case study focusing on solar PV, heat pumps and storage in a residential quarter. OR Spectrum **40**(1), 265–310 (2018)
10. T.L. Bergman, A.S. Lavine, F.P. Incropera, D.P. DeWitt, *Fundamentals of Heat and Mass Transfer*, chapter 1, 7th edn. (John Wiley and Sons, Jefferson, 2011)
11. P. Carrasqueira, M.J. Alves, C.H. Antunes, Bi-level particle swarm optimization and evolutionary algorithm approaches for residential demand response with different user profiles. Inf. Sci. **418–419**, 405–420 (2017)
12. M.J. Alves, C.H. Antunes, A semivectorial bilevel programming approach to optimize electricity dynamic time-of-use retail pricing. Comput. Oper. Res. **92**, 130–144 (2018)

Weighted Fair Queuing as a Scheduling Algorithm for Deferrable Loads in Smart Grids

Tuncer Haslak

Abstract Weighted Fair Queuing (WFQ) is implemented for the problem of load scheduling in demand side management. Power demand, wait time and group-to-group fairness are the basis for three variants of the algorithm's implementation. The results are compared to a Greedy strategy with regard to the residual of renewable power supply and the suggested measures of fairness. WFQ proves comparable to Greedy in terms of the primary objective and in addition is capable of equally distributing resources and distress caused by deferral.

Keywords Demand side management · Optimization · Weighted fair queuing

1 Introduction

Renewable energies are a resource that strain the grid through intermittent availability and difficulty in prediction. Demand side management addresses the issue by reversing the paradigm of grid operation and controlling power consumers instead of only power generators.

Within demand side management there is a need for robust algorithms that face the unpredictability aspect of renewable energies, which makes real-time algorithms with no forecast information favorable.

Finding solutions that improve the residual power problem is only the first step. Said problem is the question of how unused renewable supply should be handled. Using the *Greedy* algorithm significant improvements can be accomplished. However, the results indicate that distress would be unevenly spread in the consumer population. If distress is distributed unevenly, compensation would also be distributed unevenly, which means that the market is not design towards fairness.

T. Haslak (✉)
University of Erlangen, Erlangen, Germany

Hof University, Hof, Germany
e-mail: tuncer.haslak@fau.de

V. Bertsch et al. (eds.), *Advances in Energy System Optimization*,
Trends in Mathematics, https://doi.org/10.1007/978-3-030-32157-4_8

This paper introduces *Weighted Fair Queuing* (*WFQ*) as a scheduling algorithm for deferrable loads in smart grids. Three variant implementations are presented that feature different concepts of fairness: serving more and simultaneously smaller loads, fairly distributing wait times, and treating any number of groups equally.

2 Materials and Methods

2.1 Model

The experimental environment is designed in AnyLogic and uses an interface to control power consuming processes, as presented in [1]. The interface is suitable for any process that is capable of prolonging periods of its activity or inactivity as can be seen in Fig. 1. This requires no intimate knowledge of the process: when an algorithmic decision is made to alter the behavior of an individual load, the request is passed on to the agent representing the load. In accordance with its internal conditions it then accepts or rejects. This is to replicate the sovereignty of loads, as in actuality algorithms are informed of the current states and consider them accordingly.

Every process is divided into 4 states: activity and inactivity, and for each of those one deferrable and one non-deferrable. Processes can alter their power demand from step to step or follow a mathematical function–anything that describes the real behavior. The model captures distinct periods in which switching can be deferred.

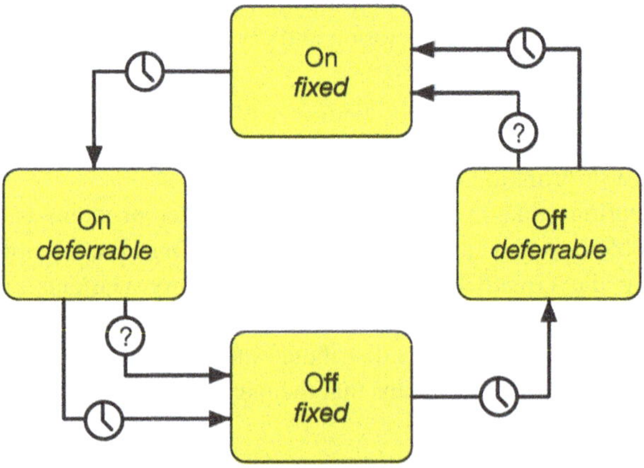

Fig. 1 State automaton that governs the state changes for deferrable loads; States can be changed by progression of time (clock) or request (question mark)

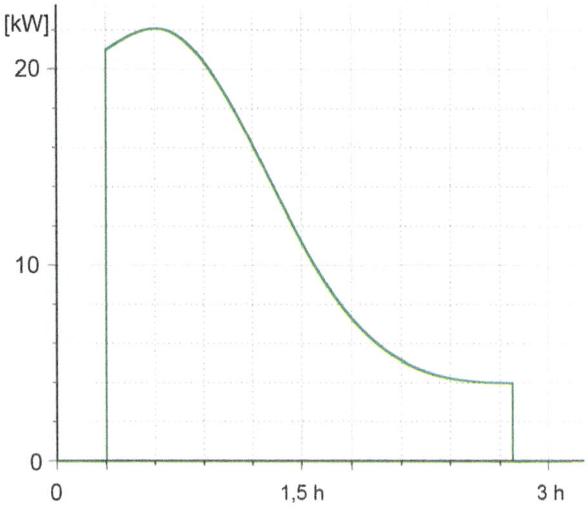

Fig. 2 This graph shows the power demand of a generic electric vehicle over approximately 2.5 h with a peak power demand of 22 kW. The simulation contains 300 consumers, this graph exemplifies one specimen

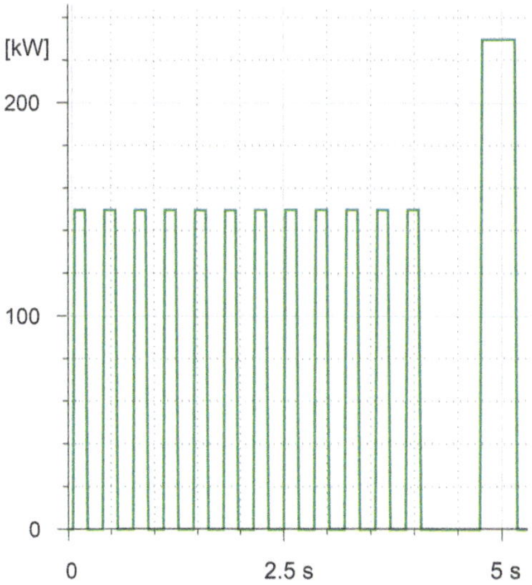

Fig. 3 The automated welding machine requires 150 kW over 1–3 s, repeated 12 times over. In the final step once 230 kW are needed for 5 s. This graph serves to show the internal working of a consumer agent

The deferral and operation times are a matter of survey in the companies (interviews, data sheets and surveillance). The total load shape is based on real loads that are replicated with the simulation. Figures 2 and 3 are two examples of the power demand and the repeated activation of loads.

The analyzed algorithms make no use of forecasting of any kind; they operate based on information of the "now". The power supply is processed data-point-by-data-point, left to right.

2.2 Population

The model population consists of approximately 300 individual loads. These encompass beverage production, metal working, glass finishing, plastics production, textile treatment, sewage treatment, electric vehicles and compressors for both pressure applications and freezing. The switching behavior of a load for the purposes of the simulation is detailed. The realistic accuracy was verified by a test implementation in [2]. The composition of the population, i.e. the proportions of types of loads are elected to reflect a midsized city by approximation, based on statistics found in [3, 4] and [5]. The overall peak power demand across the whole population is 25 MW, with single members requiring power between 1 and 350 kW. The average population power demand is 5.8 kW (cf. Fig. 7). State changes are deferrable by 15% to 50% and depend on the current and future state (arrows in Fig. 1) of the individual load. At any given time an average of approximately 20% of loads is deferrable.

2.3 Objective Function

Considering the negative impact of renewable energies, optimizing the load population toward a renewable energy supply profile seems adequate. As the volatility of the renewables is the straining property, immediate consumption by the deferrable loads would lead to the minimization of the residual power function and thereby stabilization. Residual power (R) is the amount of power leftover from renewable supply after subtracting the demand. A sample from historic supply data is picked, that consists of wind and solar energy over 24 h with a resolution of one data point per 3 s. This deviates from the resolution of simulated demand, which is continuous. Figure 4 shows the sample. At this resolution the objective function incorporates no useful gradient information.

However, selecting a random supply function is not sufficient: First the data points of the supply function must be scaled according to the demand of the population of deferrable loads. The first simulation experiment is a simple observation of how much power the consumer population requires when there is no optimization at all. This is the so called "natural run" that allows any and all requests to switch on or off. This unimpeded 24 h simulation returns a power demand function ($P(t)_{Demand}$)

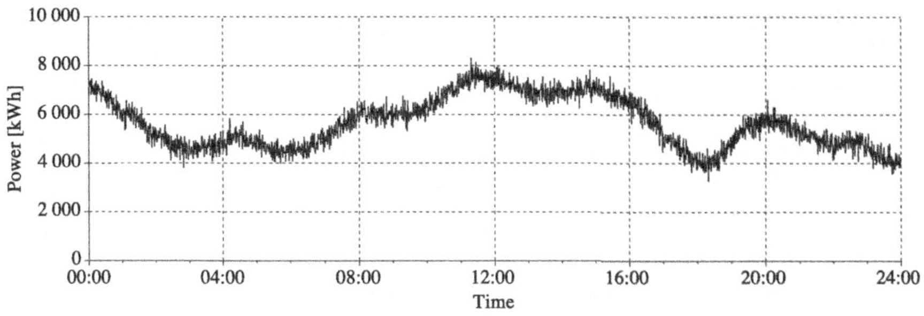

Fig. 4 Sample supply function, sum of wind and solar energy in northern Germany

in Watts. The integral thereof is the energy demand (W_{Demand}) in Watt-hours (Eq. 1).

$$\int_0^{24h} P(t)_{Demand}\, dt = W_{Demand} \tag{1}$$

In Eq. (2) the integrals of the natural run and the supply functions which are W_{Demand} and W_{Supply} are equated. This yields a result for c, the scaling factor. The result of this approach is a function of power supply ($P(t)_{Supply}$) that can satisfy the demand in terms of energy (kWh), but in terms of the progression of power over time (kW) any algorithm must manage transient deficits.

$$\begin{aligned} W_{Demand} &\stackrel{!}{=} \mathbf{c} \cdot W_{Supply} \\ P(t)_{Demand} &\neq P(t)_{Supply} \end{aligned} \tag{2}$$

This scenario seems manageable concerning not only algorithms, but is also realistic with regard to resource allocation. The latter meaning that purely renewable supply seems unrealistic and would probably feature a safety factor to account for fluctuations in demand. At the same time, however, in an economic sense a rough balance between supply and demand is sensible, because of investment and operative costs. At the least, equal supply and demand appears to be a good initial metric.

To facilitate the comparison the summation from equation (3) is used, approximating the integral of the residual power for discrete time steps:

$$R = \sum_{t=0}^{T} |P_{supply} - P_{demand}|(t) \tag{3}$$

This condenses the result of one 24 h simulation into one manageable value.

2.4 Algorithms

2.4.1 Greedy

The *Greedy* strategy (Algorithm 1) is a heuristic search algorithm. It is incomplete, but fast and very simple. As [6] shows it generally yields good results, though it is not capable of systematically finding global optima. I chose this algorithm as a starting point for its popularity, non-specificity and low computational requirements. I have previously outlined the use of the *Greedy* algorithm in [1] and [2]. The algorithm sorts the available candidates according to size and attempts to activate as many as possible, as can be seen in Algorithm 1.

2.4.2 Weighted Fair Queuing

Weighted Fair Queuing (*WFQ*) is a sophisticated algorithm designed by Demers et al. (cf. [7]), building on the work of Nagle [8]. *WFQ* manages congestion in gateway queuing of datagram networks. The problem occurs when participants in a network attempt to send data packets, but some entities always send larger packets than others. *WFQ* replaces the pragmatic FIFO (first in, first out) paradigm by categorically distinguishing datagram sources by packet size and permanently assigning them to queues with predetermined weights. Weight, packet size and a queues history determine which packet gets serviced next.

Identifiable parallels to the scheduling of deferrable loads lead to the application of *WFQ* in a novel way. The deferrable loads also feature a well-defined requirement for resources which is their power demand in Watts. This parameter is equatable to the packet size, as is the network bandwidth to the usable residual power. The significant reason that makes a more sophisticated algorithm desirable is the unfair dissemination of resource with the *Greedy* strategy. By sorting according to size, *Greedy* categorically prefers large loads, which disadvantages smaller loads.

Algorithm 1: Greedy

```
while forever do
    forall Load do
        if Residual > Load(i) then
            activate;
            calculate Residual;
        else
            deactivate;
        end
    end
end
```

WFQ (2) employs a heuristic called *virtual finish time*[1] F (Eq. 4), which is calculated using the packet size, or in this case the wattage of a load l and the weight w of each queue i. Every load is usually permanently assigned to a queue, as are weights to queues.

$$F_i = \frac{l}{w} \tag{4}$$

This yields a numeric value for each load that wants to schedule for activation. For each queue a so-called *session virtual finish time* SF_i is calculated (Eq. 5).

$$SF_i = F_i + \sum_{n=0}^{i-1} F_n \tag{5}$$

Within every queue, all elements are queued on a FIFO basis. The element of the queue i to be scheduled next is F_i. All previously successfully activated elements are added with the summation term F_n. At every point in time the algorithm selects the queue with the lowest SF_i value. The significant difference in comparison to the original *WFQ* implementation is the explicit check (Algorithm 2, ln. 10) whether the current residual power is sufficient for the regarded load. In the network realm decreasing bandwidth will not cause a packet to be rejected service–it will, however, take increasing amounts of times to submit. This "flow" property is not given in the case of deferrable loads, therefore "choking" cannot be applied, and switching decisions are discrete.

Fairness

In datagram networks fairness is the property of allotting a certain resource to any participant according to a predefined key. Such a key maybe simple and assign equal fractions to everyone. In the case of *WFQ*, however the distribution scheme accounts for a priori observations in terms of the typical resource requirements of each participant. With *WFQ* bandwidth cannot be manipulated which means that the amount of sheer data transmitted will at best remain the same (though more packets cause more gaps). In fact the objective is to service a greater number of participants.

Fairness of Wattage

This exact thought is transferred to demand side management applied to a load's wattage. In contrast to the original network application participants do not forfeit

[1]heuristic, not physical time

Algorithm 2: Weighted fair queuing

```
begin function calculateSFs
    forall queues do
        F(i) ← l/w ;
        SF(i) ← F(i) + F(n);
    end
end
while ¬Abort do
    calculate Residual;
    calculateSFs forall Load do
        if Residual > Load(i) ∧ ¬Restart then
            activate;
            Restart ← true;
        else
            Abort ← true;
        end
    end
end
```

after multiple trials in vain. It is implausible to assume that a load be denied access altogether. Loads forcing themselves to power on is unfavorable, but in line with the conventional grid operation paradigm of supplementation in such a case. As the key influenceable property is the length of part of the "on" and "off" episodes of each load, disproportionately large loads compensate with their "on" time. This means that larger loads are operated for shorter periods of time by the algorithm. The objective of employing this queuing scheme is to serve more participants in the same time interval. Here fairness is targeting decongestion.

Fairness of Distress

"Wait time" is a second, additional fairness measure worthy of investigation. Wattage is a strictly objective parameter. If we consider the period that a process can wait to be served, this requires very intimate knowledge of the process to determine the validity or honesty of time parameters as declared by a machine operator. A company might easily misrepresent this information. For the simulation experiment at hand the processes were personally surveyed and documented during live observations. Especially as there was no motivation presented to falsify data, the time parameters can considered to be sufficiently accurate.

In this regard fairness would disregard the wattage of a load and consider the time it has been waiting to get served. Fairness in this sense is the equal distribution of waiting-distress among the entirety of the load population.

For this variant fixed queue assignments are replaced with dynamic ones. The longer a process waits, the further it moves into queues that make it more likely to be picked next. To this end standardized wait time is favorable, i.e. the fraction of time waited in proportion to the maximum amount of time the load can wait (Eq. 6). Without this measure loads that provide less flexibility are irrationally rewarded.

$$t_{st} = \frac{t_{waited}}{t_{max}} \tag{6}$$

Fairness Group-by-Group

As for a third measure of fairness the load population is subdivided into regions of equal size, representative of quarters or districts. In this configuration each queue represents one region, and regions compete for the available resources. This option is designed to grant each group the same right to resources. Any attribute can be used to queue in this manner. This generic queuing scheme targets an equal distribution of deferral times and power to independent groups, but not within the groups. As the *virtual finish time* paradigm is still applied, smaller loads are preferred.

Weights and Queues

There is no definitive answer on how weights should be chosen. Existing procedures such as that outlined in [9] are not applicable because the prevailing problem of node congestion in networks is not a relevant phenomenon. Therefore a mode of selecting weights should rely on the available parameters. Each of the fairness measures suggested above requires its own weighing scheme.

2.4.3 Weight by Power (WFQ-Power)

Considering that the resource "power" is engrossed correlating with wattage, a load's wattage should inversely determine its likelihood of being elected. As a number of queues to which all loads are assigned, 5 was elected as it is typical. The fifth of the population with the highest wattage is assigned to the according quintile, and so forth. The weight for each queue is determined by computing its average wattage. Following equation (4), fairness is established when the *virtual finish times* of all individual loads are equal. This means solving for w and equating the average wattages of all queues. Table 1 shows which weights are assigned to queues, and the respective average wattage within the queue.

For example: At 250 kW load # 1 is a member of the queue weighted at 0.09, i.e. the group of highest wattages. Compared to bottom quintile member load # 2 of 20 kW and its weight of 1. Load # 1 has a virtual finish time 138 times higher. The mismatch in wattage reflects proportionately in the weights.

Table 1 Averages of power in queues and corresponding weights (*WFQ-Power*)

μ of power in queue	Queue weight
220 kW	0.09
120 kW	0.17
90 kW	0.23
40 kW	0.5
20 kW	1

Table 2 Loads for *WFQ-Time* are dynamically assigned to queues by the percentage of wait time expended from their maximum so far. In order to augment urgency, queue weights are doubled from queue to queue

Expended wait time	Queue weight
80–100%	1
60–80%	0.5
40–60%	0.25
20–40%	0.125
0–20%	0.0625

2.4.4 Weight by Wait Time (WFQ-Time)

Queuing by standardized wait time is carried out dynamically. In contrast to wattage-queuing there is no a priori assignment. With increasing wait time, a load is assigned to queues with higher weights that equate to higher likelihoods of getting picked. Once again there are five queues, each designated for a 20% interval of wait time completed. Table 2 shows how weights are governed for each queue. Decreasing weights augment urgency.

2.4.5 By Region (WFQ-Geo)

In the case of grouping by region equal weights are assigned to all queues. As the fictitious regions comprise equal portions of every category of the overall population, and given the equal weights, this queuing scheme acts comparable to a round-robin. This is true for all types of *WFQ* with equal weights (cf. [8]).

3 Model Limitations

The proposed model requires highly automated processes. The underlying work plans are machine accessible so that the ensuing steps and their duration can be projected. Semi-automation or increased human interaction would hinder this procedure.

Even though they were not encountered during survey, processes that can continuously adapt the level of their power input can only be replicated as a series of discrete steps. On the contrary this model is a response to the lack of continuously adaptable processes. Yet this is a limitation.

In addition the size of the consumer population cannot be arbitrarily increased. To this end an arbitration mechanism appears feasible. Multiple optimization algorithms would work on a fraction of a larger problem which in turn is consolidated by optimization layers higher up in the hierarchy. This approach could also help to reduce the amount of datagrams that are sent between the loads and the optimizer.

4 Results

In each simulation there is a population of approximately 300 manageable loads that must be scheduled over 24 h following a randomly selected supply signal. There are five experiments:

1. Natural run: All loads operate without any algorithmic interference. This is a reference trial.
2. Greedy
3. WFQ: Power
4. WFQ: Wait time
5. WFQ: Region

Results are compared regarding the fulfillment of the primary optimization objective from equation (3)–the minimization of residual power deviations. The three queuing schemes of the *WFQ* variants have different secondary objectives. The algorithm in experiment 3, with weights defined by power demand, aims to serve more participants, especially smaller ones. Experiment 4 entails queuing by wait time, which is designed to equally distribute the wait time percentages over the entire population. Queuing by region, experiment 5, is designed to allocate the same resources to arbitrarily defined subsets of the population.

In order to judge the fulfillment of these secondary objectives additional parameters must be analyzed. *Cycles* counts all activations which is the sum of every individual power-on that was granted–in the datagram sense this is analogous to serviced packages. The *standardized average wait time* is indicative of how long every activation was deferred in percent of the maximum. *Deferral length* is the sum of all deferrals as time. The *variance* of this value is representative of how evenly deferrals are distributed among the population.

Table 3 summarizes the results of experiments 1–4. As outlined in Sect. 2.3 the significant data is the power consumption function that every algorithm causes. Figure 5 is the consumption profile caused by the *Greedy* strategy. Consumption curves for all algorithms including the natural run (experiment 1) can be found in the appendix (Figs. 7, 8, 9, 10, and 11). The visual differences in the residual curves are **exiguous** due to the algorithmic similarity. The key difference is the residual energy. Systematic differences do not express discernibly in the graphs. Showing which load

Table 3 While *Greedy* performs best in terms of *residual*, it causes a high variance in deferral lengths (inequal waiting). *WFQ-Power* and *WFQ-Time* accomplish their secondary objectives of equal distress distribution (*variance*) and increasing the serviced loads (*cycles*) at the cost of energy optimization (*residual*)

	Natural run	Greedy	WFQ-Power	WFQ-Time	WFQ-Geo
Residual [MWh]	13,455.3	**6261.8**	6524.7	6749.9	6720.7
Avg. standardized wait [%]	0	34.0	34.7	34.2	35.9
Avg. deferral length [h]	0	3.83	3.69	3.90	3.93
Variance of deferral length	0	0.2027	0.1730	**0.0973**	0.169
Cycles (switch on & off)	577,783	419,228	**489,450**	439,278	443,366

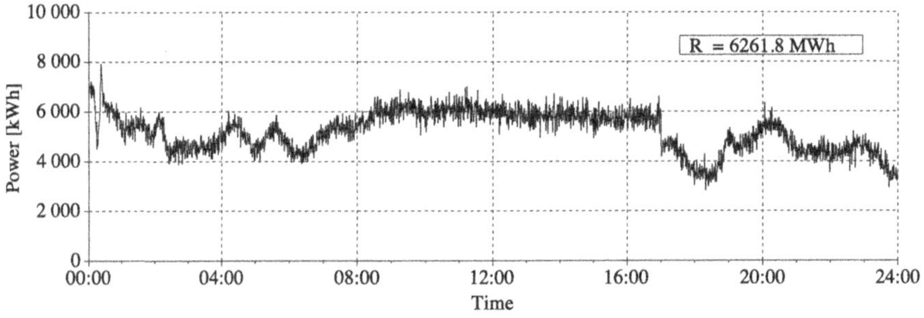

Fig. 5 Power consumption for the Greedy algorithm, residual *R* in upper right corner

Table 4 This table shows the distribution of wait times of all queues of the *WFQ-Geo* algorithm. The objective here is to treat all queues equally, which hinders global optimization objectives. Fairness reflects in the similarity of waiting times

Queue	1	2	3	4	5
Avg. standardized wait [%]	35.9	35.6	36.1	35.8	36.0
Avg. deferral length [h]	3.89	3.95	4.02	3.89	3.92
Variance of deferral length	0.172	0.169	0.173	0.168	0.164
Cycles	88,243	87,922	88,101	88,450	87,921

is activated at any given time offers no insights from algorithm to algorithm, as no pattern emerges.

The *Greedy* algorithm performs best by reducing the residual by 54.8%. *WFQ-Time* improves the deferral distribution by a factor of 2.08, while maintaining a very similar deferral length.

Table 4 shows the results of all queues of experiment 5, each representing one region. As this algorithm is specified not to improve the global population, but to serve queues equally, its results cannot be directly compared to the other experiments. The only exception is the residual. Queue deferral length is on average 11% higher compared to *WFQ-Power*. Queue by queue comparison shows that results group close together for all parameters, but worse in general.

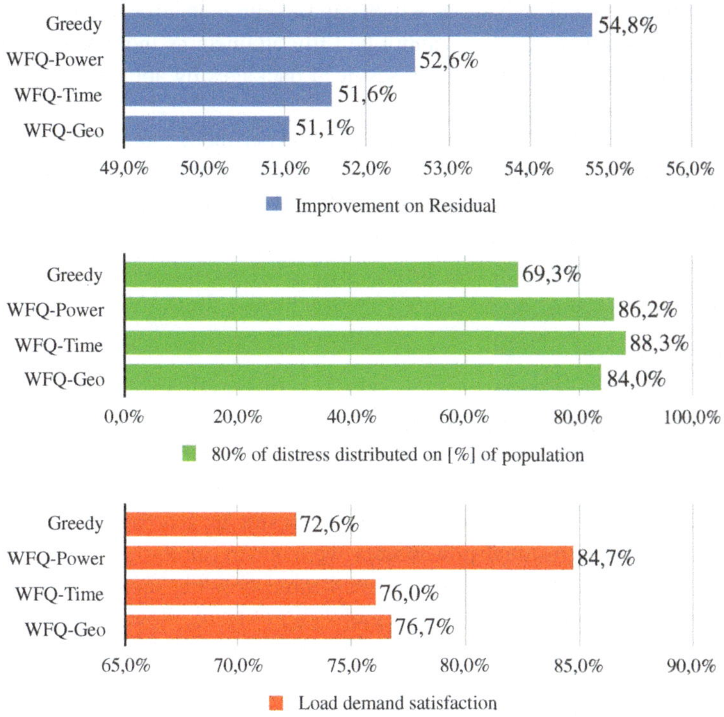

Fig. 6 The *WFQ* algorithms can compete with the versatile *Greedy* algorithm in terms of the residual optimization (blue). However *Greedy* focuses 80% of distress on only 69% of the population (green). In addition *WFQ-Power* rejects especially few loads (red)

Figure 6 summarizes the advantages of *WFQ*: it is an improvement on the *Greedy* algorithm which is fast and versatile. At the cost of slightly worse results with regard to the optimization of the power residual, *WFQ* algorithms distribute distress more equally, and serve more loads all else being equal.

5 Discussion

5.1 Improvement

As to be expected the natural run that features no algorithm or adaptation to the objective function causes a high residual. The *Greedy* algorithm delivers the greatest improvement in terms of residual. At the same time this algorithm unevenly distributes the wait time distress across the population which is expressed in the variance of the average deferral time. *WFQ-Time* significantly narrows the variance of deferral, following its secondary objective. The *WFQ-Power* variant is capable of placing more individual activations which can be seen in its *cycle* count. Keeping in mind that the objective function cannot be fully attained, as indicated in Sect. 2.3, the algorithm still reduces the gap in serviced packages by approximately 60% in comparison to *Greedy*.

The results from the algorithm *WFQ-Geo* in experiment 5 are inferior in general as it causes longer wait times and the variance thereof indicates that wait times are heterogeneously distributed within queues. The objective of equal queue treatment is achieved at the cost of worse results all in all.

5.2 Distress

Distress distribution is a desirable trait in demand side management that is not trivially deducible. Wait time is a parameter that cannot be easily subverted, if for example the maximum deferral time is repaid as an incentive. Weighted Fair Queuing is capable of delivering comparable result to a *Greedy* strategy in the primary objective of residual power improvement. In addition it can be adapted to pursue additional objectives. I presented three possible measures of fairness: equality between multiple queues or regions (*WFQ-Geo*), the even distribution of wait times (*WFQ-Time*), and serving more participants (*WFQ-Power*)–the latter being most similar to the original algorithm. This shows that *WFQ* can be used to introduce fairness to the selection of deferrable loads.

5.3 Methodological Contribution & Policy Advice

Despite its complexity *Weighted Fair Queuing* is a staple of network congestion management. It is highly functioning and ubiquitous because of its performance and its ability to accomplish fairness. The application to energy distribution problems is novel. At the same time it introduces the problem of fairness into the matter, which was disregarded because of the focus on residual power optimization, monetary compensation or welfare gain.

The optimization problem presented and solved in this paper is significant in its size regarding multiple dimensions. Firstly the time resolution exceeds the commonly quoted 15 min intervals that stem from the tertiary grid balancing realm. Grid stabilization however can only be achieved when the shrinking number of spinning masses can be replaced. The realization that compensatory measure must move below the 30-seconds-threshold is at the core of this research. Secondly, the population is substantial with 300 members and 25 MW peak. Each load is replicated as an agent with internal processes, variables and decision making. Furthermore their states are not estimated or stochastically approximated–decision making and transmission is acute, meaning that statuses are inquired, evaluated, decided on and requests dispatched. Thirdly, no simplifying assumptions are applied to the target function (power supply) such as smoothing. The complexity of the problem is embraced to the extent that the resolution of the objective function (supply) is the restricting factor. The argument is in favor of preparedness, as methodologically the question answered here is that demand side management features the necessary scalability and agility.

The suggested algorithms are implementations of centralized algorithms. As the technology is still at an early stage centralization is acceptable. Especially as the methodological exploration of this topic–which this research is part of–is still ongoing. For future applications, however, this is not advisable, as a single entity will possess all information on all clients (loads). The controlling entity would be vulnerable to attacks . The presented model, is acting on a request basis which means that a load can reject any suggestion to change its state. Designing intrinsically safe systems increases complexity, but is quintessential policy advice.

5.4 Practical Implications

For companies owning deferrable loads the practical implications are low, especially if processes are highly automated. The test implementation presented in [1] required no manual interaction. While loads were in interference the optimization paused and resumed automatically. The more a deferrable load is interwoven into the process of a company, the more difficulty can arise from automated unexpected deferrals. Deferrable loads are an effort that aim to improve the usage of renewable loads. Some resulting distress is to be expected. Although not insinuated here, market design should compensate for this.

The suggested interface is at the core of the research presented, as it allows for the load to negotiate deferral and reject it. From a load's perspective *WFQ-Time* is the most favorable, as it guarantees that no load is overburdened with deferral. The *Greedy* algorithm is unfavorable, as some loads are asked to defer significantly more often than others (unfairness).

5.5 Algorithm Comparison

There is a multitude of optimization algorithms that can be used. The advantage of the above presented algorithms is their simplicity. They can be categorized as local search algorithms, which means that they use a type of heuristic, like SF_i in Eq. (5). As there are no simplifying assumptions, and every load is designed to be addressed with a direct request, the problem presents itself to be massive and without useful gradient information, which excludes classic optimization algorithms.

A suitable alternative would for example be "*Simulated Annealing*" (cf. [10]). It is an algorithm that begins by amply exploring the solution space with random solutions. A solution in this case is an allowable sequence of activation and pause for all loads. The algorithm continues by rejecting or accepting results, based on the metaheuristic property ("temperature"). In principle this algorithm converges on the global optimum solution, which local search algorithms are not systematically capable of. The significant disadvantage is considerably higher run time. A preliminary test implementation shows that, after 14 h *Simulated Annealing* delivers results, which are comparable to the *Greedy* strategy which takes approximately 15 min on the same desktop computer (2.5 GHz Intel Core i7 with 16 GB 1600 MHz DDR3 memory).

6 Related Work

Gellings laid out the fundamentals of demand side management in [11].

The objective function and its definition are based on [12] and [13].

Weighted Fair Queuing and the approach of *Generalized Processor Sharing* are based on [7, 8] and [14]. These original sources were used for the adaptation to demand side management.

Most related work is based on [15] where an optimal deferrable load control problem is defined. The focus is placed on market parameters which is why price bounds are employed. [16] focuses on the problem laid out in this publication. To solve this problem a decentralized algorithm is employed, which communicates the power residual to all participants. This is in contrast to centralized solution efforts, but capable of scheduling multiple instances of the same load. [17] utilizes a similar approach but only for singular placement of loads in 24 h.

[18] uses a stochastic model which defers one activation of each load and no reactivation. The underlying idea is very strongly connected to the paradigm of grid operation. This means that the time scales of balancing power in the European grid are adopted and all activity takes places in time steps of 15 min, and deferral times are at least 4 h. The significant difference is the lack of understanding and anticipating load behavior. The publication focuses on the simulation of wind power only and the cost of demand side management.

[19] has a similar approach in that only one activation instance of every participant is considered. The multitude of loads is approached as a combinatorial problem. Here the smallest time step is one hour. In neither of these loads can extend their operative time.

Formulated as a constraint problem regarding the availability of the demand side, [20] manages a day-ahead scenario by interacting with energy markets. By the introduction of welfare they move away from monetary quantification. They outline the difference in time scale between markets (larger steps) and the load behavior (smaller steps).

Appendix

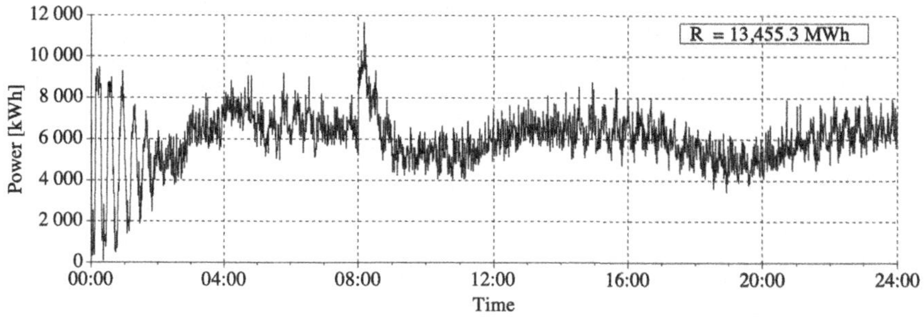

Fig. 7 Power consumption for a non-optimized reference run; 1 h equals 100 simulation units

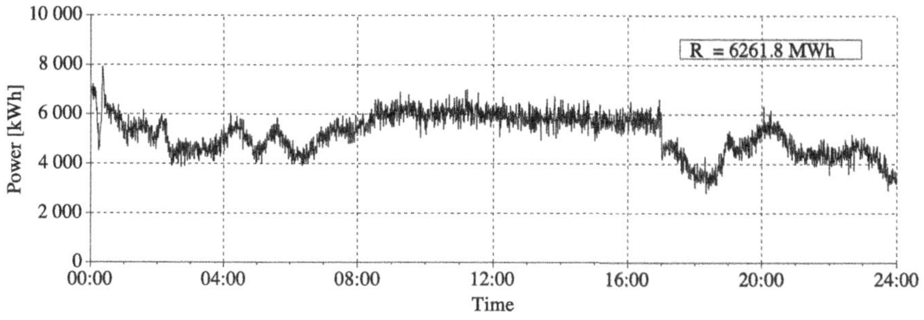

Fig. 8 Power consumption for the Greedy algorithm

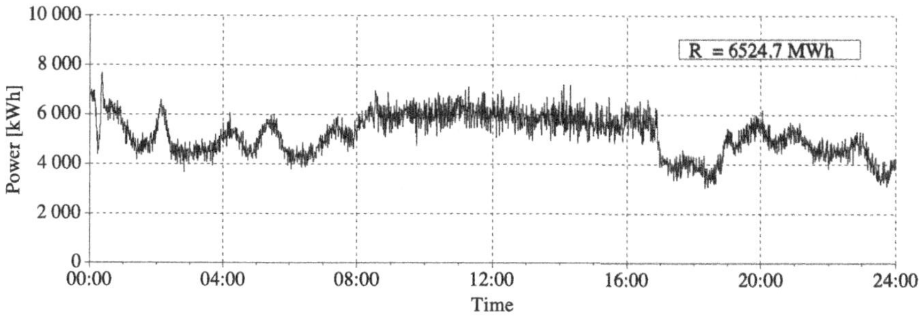

Fig. 9 Power consumption for the *WFQ-Power* algorithm

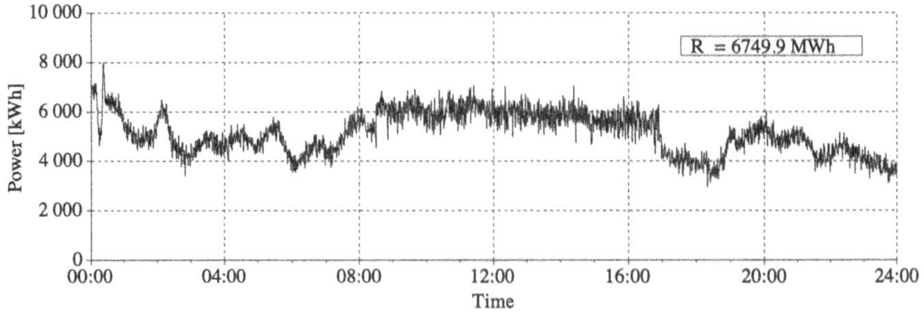

Fig. 10 Power consumption for the *WFQ-Time* algorithm

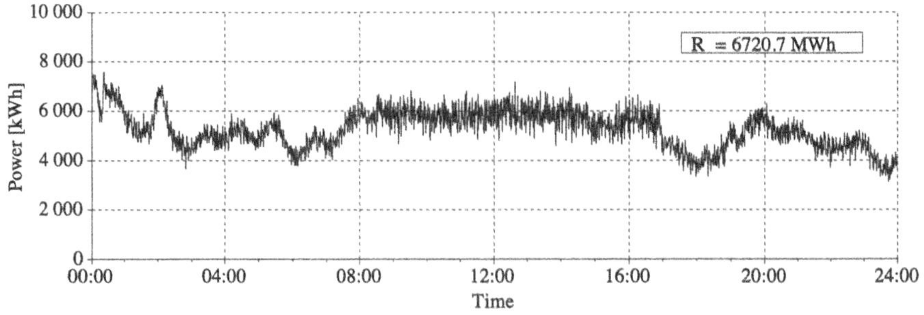

Fig. 11 Power consumption for the *WFQ-Geo* algorithm

References

1. T. Haslak, *SACI 2016 – 11th IEEE International Symposium on Applied Computational Intelligence and Informatics, Proceedings*, 2016, pp. 381–384. https://doi.org/10.1109/SACI.2016.7507406
2. C. Brabec, M. Neswal, R. German, Final Project Report: Smart Grid Solar. Tech. rep. (2019)
3. A. Kollmann, C. Amann, C. Elbe, V. Heinisch, Lastverschiebung in Haushalten, Industrie, Gewerbe und kommunaler Infrastruktur Potenzialanalyse fuer Smart Grids. Tech. Rep. 2 (2013)
4. M. Buddeke, C. Krüger, F. Merten, Modellbeschreibung: Einsatzmodell für Flexibilitätsoptionen im europäischen Stromsystem Projektbericht. Tech. rep. (2016)
5. L. von Bremen, M. Buddeke, D. Heinemann, A. Kies, D. Kleinhans, C. Krüger, F. Merten, M. Preute, S. Samadi, T. Vogt, W. Lukas, Ergebnisse und Empfehlungen des BMBF-Forschungsprojektes Regenerative Stromversorgung & Speicherbedarf in 2050. Tech. Rep. September (2016)
6. T. Weise, *Global Optimization Algorithms – Theory and Application* (2008). https://doi.org/10.1109/HIS.2007.11
7. A. Demers, S. Keshav, S. Shenker, Analysis and simulation of a fair queueing algorithm (1989). https://doi.org/10.1145/75247.75248. http://portal.acm.org/citation.cfm?doid=75247.75248
8. J. Nagle, SIGCOMM Comput. Commun. Rev. **14**(4), 61 (1984). http://doi.acm.org/10.1145/1024908.1024910
9. E. Magaña, D. Morató, P. Varaiya, in *10th International Conference on Telecommunications, ICT 2003*, vol. 2, 2003, pp. 917–922. https://doi.org/10.1109/ICTEL.2003.1191562
10. S. Kirkpatrick, C.D.J. Gelatt, M.P. Vecchi, Science **220pp**(4598), 671 (1983)

11. C. Gellings, Proc. IEEE **73**(10), 1468 (1985). https://doi.org/10.1109/PROC.1985.13318. https://ieeexplore.ieee.org/document/1457586/
12. G. Strbac, Energy Policy **36**(12), 4419 (2008). https://doi.org/10.1016/j.enpol.2008.09.030
13. W.P. Schill, Energy Policy **73**, 65 (2014). https://doi.org/10.1016/j.enpol.2014.05.032
14. A.K. Parekh, R.G. Gallager, IEEE/ACM Trans. Netw. **1**(3), 344 (1993). https://doi.org/10.1109/90.234856
15. A.J. Conejo, J.M. Morales, L. Baringo, IEEE Trans. Smart Grid **1**(3), 236 (2010). https://doi.org/10.1109/TSG.2010.2078843
16. L. Gan, A. Wierman, U. Topcu, N. Chen, S.H. Low, *Proceedings of the Fourth International Conference on Future Energy Systems (e-Energy '13)*, pp. 113–124 (2013). https://doi.org/10.1145/2567529.2567553. http://dl.acm.org/citation.cfm?id=2487179
17. G. Graditi, M.L. Di Silvestre, R. Gallea, E.R. Sanseverino, IEEE Trans. Ind. Inf. **11**(1), 271 (2015). https://doi.org/10.1109/TII.2014.2331000
18. M. Klobasa, Dynamische Simulation eines Lastmanagements und Integration von Windenergie in ein Elektrizitätsnetz. Ph.D. thesis (2007). https://doi.org/10.3929/ethz-a-005484330
19. T.P.I. Ahamed, S.D. Maqbool, E.a. Al-Ammar, N. Malik, *2011 2nd IEEE PES International Conference and Exhibition on Innovative Smart Grid Technologies*, 2011, pp. 1–4. https://doi.org/10.1109/ISGTEurope.2011.6162637
20. L. Jiang, S. Low, *2011 49th Annual Allerton Conference on Communication, Control, and Computing*, Allerton, 2011, pp. 1334–1341. https://doi.org/10.1109/Allerton.2011.6120322

Part IV
Planning and Operation of Distribution Grids

Cost Optimal Design of Zero Emission Neighborhoods' (ZENs) Energy System

Model Presentation and Case Study on Evenstad

Dimitri Pinel, Magnus Korpås, and Karen B. Lindberg

Abstract Zero Emission Neighborhoods (ZEN) is a concept studied in particular in the research center on ZEN in smart cities in Norway to reduce the CO_2 emissions of neighborhoods. One question coming along this concept is how to design the energy system of such neighborhoods to fit the ZEN definition[1]. From this definition we extract the CO_2 balance, requiring an annual net zero emission of CO_2 in the lifetime of the neighborhood. This paper proposes a MILP model for obtaining cost optimal design of ZEN's energy system and demonstrates it on a case study. Different technologies are included as investment options and, notably PV as a mean of producing electricity on-site. Wind turbines are not included in this study because they would not be suitable in the context of most cities. The results highlight the importance of PV investment in reaching the ZEN requirements. For example, around 850 kW of solar is needed for our test cases of 10,000 m^2 of floor area, for an annual energy demand of around 700 MWh of electricity and 620 MWh of heat. The investments in other technologies are small in comparison.

Keywords ZEN · Sustainable neighborhoods · Zero emission Neighborhoods · Energy system · CO_2 emissions · Optimization

1 Introduction

A ZEN is a neighborhood that has a net zero emission of CO_2 over its lifetime. Many aspects are embedded in the idea of ZEN. Energy efficiency, materials, users behavior, energy system integration are all aspects that need to be accounted for

D. Pinel (✉) · M. Korpås
Deparment of Electric Power Engineering, Norwegian University of Science and Technology, Trondheim, Norway
e-mail: dimitri.q.a.pinel@ntnu.no; magnus.korpas@ntnu.no

K. B. Lindberg
SINTEF Community, Oslo, Norway
e-mail: karen.lindberg@sintef.no

V. Bertsch et al. (eds.), *Advances in Energy System Optimization*, Trends in Mathematics, https://doi.org/10.1007/978-3-030-32157-4_9

in this concept. In addition, different parts of the life cycle can be included but in this paper we only consider the operation phase and no embedded emissions. Two types of action exist to make neighborhoods more sustainable. One is to act on the demand, via better insulation, user behavior or other efficiency measures. The other is to act on the supply and have a local energy system minimizing the CO_2 emissions. There is consequently a need for a way of designing the energy system of such neighborhoods. The questions to be answered are, which technologies are needed to satisfy the demand of heat and electricity of a neighborhood, and how much of it should be installed so that it is as inexpensive as possible. The problem is then to minimize the cost of investment and operation in the energy system of a neighborhood so that it fulfills the ZEN criteria. This paper presents an optimization model to solve such problems with a focus on operations research methodology.

2 State of the Art and Contribution

The ZEN concept is specific to this particular project, however similar topics have been studied in different settings either at the neighborhood level, the city level or the building level, for example during the research center on Zero Emission Building. In this context, K B Lindberg studied the investment in Zero Carbon Buildings [2] and Zero Energy Buildings [3] which are variations around the concept of Zero Emission Buildings. In both papers an optimization based approach is used to study the impact of different constraints on the resulting design. The second one [3] in particular uses binary variables to have a more realistic representation of the operation part (part load limitation and import/export). In [4], Gabrielli et al. tackle the problem of investment and operation of a neighborhood system and show an approach allowing to model the system complexity while keeping a low number of binary variables. It also constrains the total CO_2 emissions. It uses design days and proposes two methods for allowing to model seasonal storages while keeping the model complexity and reducing the run time. In [5], Hawkes and Leach look at the design and unit commitment of generators and storage in a microgrid context using 12 representative days per season in a linear program. It is particular in that it defines how much the microgrid would be required to operate islanded from the main grid and include this in the optimization. It also discusses the problematic of market models within microgrids. In [6], Weber and Shah present a mixed integer linear programming tool to invest and operate a district with a focus on cost, carbon emission and resilience of supply. A specificity of this tool is that it also designs the layout of the heat distribution network taking into account the needs of the buildings and the layout of each area. It uses the example of a town in the United Kingdom for its case study. In [7], Mehleri et al. study the optimal design of distributed energy generation in the case of small neighborhoods and test the proposed solution on a Greek case. Emphasis is put on the different layouts of the decentralized heating network. In [8], Schwarz et al. present a model to optimize the investment and the energy system of a residential quarter, using a two stage stochastic MILP. It emphasizes on how it tackles the stochasticity of the problem in the different stages, from raw data to the input of the optimization, and on the computational performance and scalability of the proposed method. In [9], he also studies the impact of different grid tariffs on the design of the system and on the self-consumption of the PV production. In [10], Li et al. separate the investment and the operation into a master and a follower problem. The master problem uses a genetic algorithm to find the optimal investment while a MILP is used to find the operation in the follower problem. In [11], Wang et al. also use a genetic algorithm, but at the building level and using a multi objective approach focused on environmental considerations. A life cycle analysis methodology as well as exergy consumption are used to assess the design alternatives. In [12], Mashayekh et al. uses a MILP for sizing and placement of distributed generation using a MILP approach including linearized AC-power flow equations. In [13], Yang et al. also use a MILP approach for the placement and sizing problem but consider discrete investment in technologies at the district scale. These papers give us an overview of different methods for optimal investment in the energy system of neighborhoods or buildings, but none apply the ZEN concept and the influence of tight requirements on the CO_2 emissions on the modelling and on the results has not been demonstrated.

In this paper, the focus is put on getting a fast yet precise solution that can take long term trends, such as cost reduction of technologies or climate. To this end, the proposed model uses a full year representation, ensuring a correct representation of seasonal storage of heat and electricity, and allows to divide the lifetime of the neighborhood into several periods, each represented by one year. It is also different by using the Zero Emission framework on a neighborhood level as a guide for the emission reduction constraint. This adds an integral constraints coupling each timestep and increasing the complexity of the problem. The use of binary variables is limited to the minimum.

3 ZENIT Model Description

ZENIT stands for Zero Emission Neighborhoods Investment Tool. It is a linear optimization program written in Python and using Gurobi as a solver. It minimizes the cost of investing and operating the energy system of a ZEN using periods, with a representative year in each period. Different technologies are available, both for heat and for electricity. It is most suited for greenfield investment planning but can also take into account an existing energy system. The objective function is presented below:

$$\sum_i C_i^{disc} \cdot x_i + b_{hg} \cdot C_{hg} + \frac{1}{\varepsilon_{r,D}^{tot}} \sum_i C_i^{maint} \cdot x_i$$
$$+ \sum_p \varepsilon_{r,p} \Bigg(\sum_t \Big(\sum_f f_{f,t,p} \cdot P_{f,p}^{fuel} + (P_{t,p}^{spot} + P^{grid} \tag{1}$$
$$+ P^{ret}) \cdot (y_{t,p}^{imp} + \sum_{est} y_{t,p,est}^{gb\_imp}) - P_{t,p}^{spot} \cdot y_{t,p}^{exp} \Big) \Bigg)$$

The objective is to minimize the cost of investing in the energy system as well as its operation cost.

The operation phase can be separated in different periods during the lifetime of the neighborhood, and one year with hourly time-steps is used for each period. In addition to technologies producing heat or electricity, there is also the possibility to invest in a heating grid represented by the binary b_{hg} that also gives access to another set of technologies that would be inappropriate at the building level. In the equation above, the $\mathcal{E}$ represent discount factors either global for the whole study (3) or for each period (2). They are calculated in the following way:

$$\varepsilon_{r,D}^{tot} = \frac{r}{1-(1+r)^{-D}} \tag{2}$$

$$\varepsilon_{r,p} = \frac{(1+r)^{-p \cdot YR}}{\frac{r}{1-(1+r)^{-YR}}} \tag{3}$$

The calculation assumes that reinvestment in this technology is made for the whole lifetime of the neighborhood, and is discounted to year 0. The salvage value is also accounted for. The formula used is:

$$C_i^{disc} = \left(\sum_{n=0}^{N_i - 1} C_i^{inv} \cdot (1+r)^{(-n \cdot L_i)} \right)$$
$$- \frac{N_i \cdot L_i - D}{L_i} \cdot C_i^{inv} \cdot (1+r)^{-D} \tag{4}$$

$$with : N_i = \left\lceil \frac{D}{L_i} \right\rceil \tag{5}$$

In the objective function, $y_{t,p}^{exp}$ represent the total export from the neighborhood. It is simply the sum of all exports from the neighborhood: $\forall t, p$

$$y_{t,p}^{exp} = \sum_g y_{t,p,g}^{exp} + \sum_{est} (y_{t,p,est}^{gb\_exp} + y_{t,p,est}^{pb\_exp}) \cdot \eta_{est} \tag{6}$$

The most important constraint, and what makes the specificity of the "Zero Emission" concept, is the CO_2 balance constraint. It is a net zero emission constraint of CO_2 over a year. It takes into account the emissions from the used fuels and electricity with the corresponding CO_2 factors for the emission part and the exports of electricity for the compensation part. In this study the same factor is used for imports and for exports of electricity. This constraint is expressed below, $\forall p$:

$$\begin{aligned}
&\sum_t ((y_{t,p}^{imp} + \sum_{est} y_{t,p,est}^{gb_imp}) \cdot \varphi_e^{CO_2}) \\
&\quad + \sum_t \sum_f (\varphi_f^{CO_2} \cdot f_{f,t,p}) \leq \sum_t (\sum_{est} (y_{t,p,est}^{gb_exp} \\
&\quad + y_{t,p,est}^{pb_exp}) \cdot \eta_{est} + \sum_g y_{t,p,g}^{exp}) \cdot \varphi_e^{CO_2}
\end{aligned} \tag{7}$$

In the particular ZEN framework of this study, the idea behind the compensation is that the electricity exported to the national grid from on-site renewable sources allows to reduce the national production, and thus to prevent some emissions from happening. The corresponding savings, the compensation, stand on the right-hand side of the equation. In the ZEN framework, this constraint is set as an annual constraint. It can however also be used for shorter periods of time.

Other necessary constraints are the different electricity and heat balances which guarantee that the different loads are served at all times. The electricity balance is represented graphically in Fig. 1. The corresponding equations are also written below. The electricity balance is particular because, we want to keep track of the origin of the electricity sent to the battery. It is managed by representing each battery

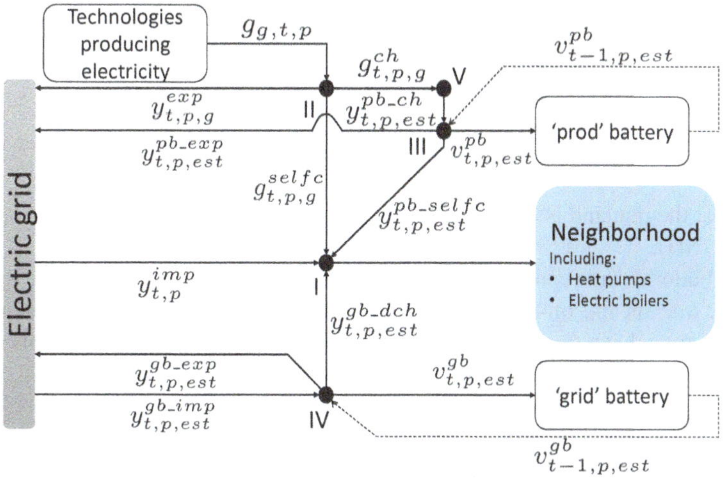

Fig. 1 Graphical representation of the electricity balance in the optimization

as a combination of two other batteries: one is linked to the on-site production technologies, while the other is connected to the grid. It allows to keep track of the self-consumption and to differentiate between the origin of the energy for the CO_2 balance.

Node I (8) represents the main electric balance equation while II (9) and V (10) are only related to the on-site production of electricity. Node II (9) describes that the electricity produced on-site is either sold to the grid, used directly or stored, while node V (10) states that at a given time step what is stored in the batteries is equal to what is in excess from the on-site production.

Electricity balance I: $\forall t, p$

$$\begin{aligned} y_{t,p}^{imp} + \sum_{est}(y_{t,p,est}^{gb\_dch} + y_{t,p,est}^{pb\_selfc}) \cdot \eta_{est} + \sum_{g} g_{g,t,p}^{selfc} \\ = \sum_{e} d_{e,t,p} + \sum_{b}\sum_{hp} d_{hp,t,p,b} + \sum_{b} E_{b,t,p} \cdot A_b \end{aligned} \tag{8}$$

Electricity balance II: $\forall t, p, g$

$$g_{g,t,p} = y_{t,p,g}^{exp} + g_{g,t,p}^{selfc} + g_{t,p,g}^{ch} \tag{9}$$

Electricity balance V: $\forall t, p$

$$\sum_{g} g_{t,p,g}^{ch} = \sum_{est} y_{t,p,est}^{pb\_ch} \tag{10}$$

Heat also has its own balance, that guarantees that the demand of each building is met:

$$\begin{aligned} \sum_{\gamma \in \mathcal{Q} \setminus \mathcal{HP}} q_{\gamma,t,p} + \sum_{b}\sum_{hp} q_{hp,t,p,b} \\ + \sum_{hst} \eta_{hst} \cdot q_{t,p,hst}^{dch} = \sum_{b} H_{b,t,p} \cdot A_b + q_{t,p}^{ch} \end{aligned} \tag{11}$$

Note that the demand is not divided between domestic hot water (DHW) and space heating (SH).

The batteries are represented, as mentioned earlier and as seen on Fig. 1, as two entities: one on the on-site production side and the other on the grid side. This means that we have two "virtual" batteries with their own set of constraints as well as constraints linking the two.

The first constraint is a "reservoir" type of constraint and it represents the energy stored in the battery at each time-step: $\forall t \in \mathcal{T}^*, p, est$

$$v_{t,p,est}^{pb} = v_{t-1,p,est}^{pb} + \eta_{est} \cdot y_{t-1,p,est}^{pb\_ch} - y_{t-1,p,est}^{pb\_exp} - y_{t-1,p,est}^{pb\_selfc} \tag{12}$$

$$v_{t,p,est}^{gb} = v_{t-1,p,est}^{gb} + \eta_{est} \cdot y_{t-1,p,est}^{gb\_imp} - y_{t-1,p,est}^{gb\_exp} - y_{t-1,p,est}^{gb\_dch} \tag{13}$$

Equations (14), (16) and (17) link both batteries. They make sure the sum of the stored energy in the "virtual" batteries is less than the installed capacity, and making sure the rate of charge and discharge of the battery is not violated. $\forall t, p, est$

$$v_{t,p,est}^{pb} + v_{t,p,est}^{gb} \leq v_{t,p,est}^{bat} \tag{14}$$

$$v_{t,p,est}^{bat} \leq x_{bat,est} \tag{15}$$

$$y_{t,p,est}^{pb\_ch} + y_{t,p,est}^{gb\_imp} \leq \dot{Y}_{max,est}^{bat} \tag{16}$$

$$y_{t,p,est}^{gb\_dch} + y_{t,p,est}^{gb\_exp} \leq \dot{Y}_{max,est}^{bat} \tag{17}$$

The storage level at the beginning and the end of the periods should be equal. $\forall p, est$

$$v_{start,p,est}^{bat} = v_{end,p,est}^{bat} \tag{18}$$

The heat storage technologies also have the same kind of equations as the batteries, for example: $\forall t \in \mathcal{T}^*, p, hst$

$$v_{t,p,hst}^{heatstor} = v_{t-1,p,hst}^{heatstor} + \eta_{hst}^{heatstor} \cdot q_{t,p,hst}^{ch} - q_{t,p,hst}^{dch} \tag{19}$$

Equations (14) to (18) also have equivalents for the heat storages. However the heat storages are not separated in two virtual entities since there is no export of heat from the building.

The power exchanges with the grid are limited depending on the size of the connection: $\forall t, p$

$$(y_{t,p}^{imp} + y_{t,p}^{exp} + \sum_{est} y_{t,p,est}^{grid\_imp,bat}) \leq GC \tag{20}$$

In order to not add additional variables, the mutual exclusivity of import and export is not explicitly stated. It is still met however due to the price difference associated with importing and exporting electricity.

In addition to the above equations, different constraints are used to represent the different technologies included. The maximum investment possible is limited for each technology. $\forall i$:

$$x_i \leq X_i^{max} \tag{21}$$

The amount of heat or electricity produced is also limited by the installed capacity:

$$\forall q, t, p : q_{q,t,p} \leq x_q \tag{22}$$

$$\forall g, t, p : g_{g,t,p} \leq x_g \tag{23}$$

The amount of fuel used depends on the amount of energy provided and on the efficiency of the technology: respectively $\forall \gamma \in \mathcal{F} \cap \mathcal{Q}, p, t$ and $\forall \gamma \in \mathcal{E} \cap \mathcal{Q}, p, t$

$$f_{\gamma,t,p} = \frac{q_{\gamma,t,p}}{\eta_\gamma} \tag{24}$$

$$d_{\gamma,t,p} = \frac{q_{\gamma,t,p}}{\eta_\gamma} \tag{25}$$

For CHPs technologies, the Heat to Power ratio is used to set the production of electricity based on the production of heat. $\forall t, p$

$$g_{CHP,t,p} = \frac{q_{CHP,t,p}}{\alpha_{CHP}} \tag{26}$$

For the heat pumps, the electricity consumption is based on the coefficient of performance (COP).

$\forall hp, b, t, p$

$$d_{hp,b,t,p} = \frac{q_{hp,b,t,p}}{COP_{hp,b,t,p}} \tag{27}$$

The heat pumps are treated differently from the other technologies because they are not aggregated for the whole neighborhood but are separated for each building. This is because the COP depends on the temperature to supply, which is different in passive buildings and in older buildings and which is also different for DHW and for SH, and dependent on the temperature of the source. The source is either the ground or the ambient air depending on the type of heat pump. The COP is then calculated using a second order polynomial regression of manufacturers data [3] and the temperature of the source and of the outside timeseries. The possibility to invest in insulation to reduce the demand and improve the COP of heat pumps is not considered. The global COP is calculated as the weighted average of the COP for DHW and SH.

The solar technologies, solar thermal and PV, also have their own set of specific constraints. $\forall t, p$:

$$g_{t,p}^{PV} + g_{t,p}^{curt} = \eta_{t,p}^{PV} \cdot x_{PV} \cdot IRR_{t,p} \tag{28}$$

$$q_{t,p}^{ST} = x_{ST} \cdot \frac{IRR_{t,p}}{G_{stc}} \tag{29}$$

The hourly efficiency of the PV system is calculated based on [14], and accounts for the outside temperature and the irradiance. This irradiance on a tilted surface is derived from the irradiance on a horizontal plane that is most often available from measurements sites by using the geometrical properties of the system: azimuth and elevation of the sun and tilt angle and orientation of the panels.

The irradiance on the horizontal plane data comes from ground measurements from a station close to the studied neighborhood which can for example be obtained from Agrometeorology Norway.[1] The elevation and azimuth of the sun is retrieved from an online tool.[2] This calculation takes into account the tilt of the solar panel and its orientation. Several assumptions were necessary to use this formula. Indeed, the solar irradiance is made up of a direct and a diffuse part and only the direct part of the irradiance is affected by the tilt and orientation. However there is no good source of irradiance data that provides a distinct measurement for direct and diffuse parts in Norway as far as the authors know. Thus we make assumptions that allow us to use the complete irradiance in the formula. We assume that most of the irradiance is direct during the day and that most is diffuse when the sun is below a certain elevation or certain azimuths. This assumption gives a good representation of the morning irradiances while still accounting for the tilt and orientation of the panel during the day. On the other hand, this representation overestimates the irradiance during cloudy days, when it is mostly indirect irradiance. Obtaining direct and diffuse irradiance data would solve this problem.

4 Implementation

The model presented in the previous section has been implemented in the case of campus Evenstad, which is a pilot project in the ZEN research center [15]. This implementation of the model and the parameters used are presented in this section. Campus Evenstad is a university college located in southern Norway and is made up of around 12 buildings for a total of about 10,000 m^2. Most of the buildings were built between 1960 and 1990 but others stand out. In particular two small buildings were built in the nineteenth century and the campus also features two

[1] lmt.nibio.no

[2] Sun Earth Tools: https://www.sunearthtools.com/dp/tools/pos_sun.php

Table 1 Technologies used in the Evenstad case and their main parameters

Technology	Inv. Cost (€/kW)	Life-time (Years)	Efficiency (%)
Building			
PV	1600	25	18
Solar Thermal	700	25	70
Air source HP	556	15	COP_t
Ground source HP	444	15	COP_t
Biomass Boiler	350	20	85
Electric Boiler	750	30	100
Gas Boiler	120	25	95
Neighborhood			
Gas CHP	739	25	45_{th}; 35_{el}
Biomass CHP	3300	25	40_{th}; 25_{el}
Heat Pump	660	25	COP_t
Electric Boiler	150	20	100
Gas Boiler	60	25	95

recent buildings with passive standards. The campus was already a pilot project in the previous ZEB center and one of those buildings was built as a Zero Emission Building. In addition, on the heating side a 100 kW CHP plant (40 kW electric) and a 350 kW Bio Boiler both using wood chips were installed along with 100 m^2 of solar collectors, 10,000 L of storage tank, 11,600 L of buffer tank and a heating grid. On the electric side, the same CHP is contributing to the on-site generation as well as a 60 kW photovoltaic system. A battery system is already planned to be built accounting for between 200 and 300 kWh. Based on this we assume in the study an existing capacity of 250 kWh. We keep those technology in the energy system of the neighborhood for one part of the study. In addition, the heating grid is kept in all cases ($b_{hg} = 1$).

The technologies included in the study are listed in Table 1 along with the appropriate parameters.

Two main sources for the parameters and cost of the technologies are used as references for the study. Most of the technologies' data is based on a report made by the Danish TSO energinet and the Danish Energy Agency [16] on technology data for energy plants. The other source includes the technology data sheets made by IEA ETSAP [17] and is used in particular for the gas and the biomass CHP. The cost of PV is based on a report from IRENA [18]. The two efficiencies reported for the CHP plants correspond to the thermal and electrical efficiency, noted by a subscript ($_{th}$ for thermal and $_{el}$ for electrical). Note that: at the neighborhood level, only ground source heat pump is considered (Table 2) and that PV is only considered at the building level but the roof area limit to the size of the PV is not implemented.

The heat storage values are based on a data sheet by ETSAP [17] while the values used for the batteries are based on a report from IRENA [19].

Table 2 Storage technologies used in the Evenstad case and their main parameters

Technology	Inv. Cost (€/kWh)	Lifetime (Years)	Efficiency (%)
Battery	350	15	94
Heat Storage	75	20	95

Table 3 Fuel cost and CO_2 factors

Fuel	Cost (€/kWh)	CO_2 Factor (gCO_2/kWh)
Gas	0.055	277
Biomass	0.041	7
Electricity	$P^{spot}_{t,p}$	17

The values in Table 3 come from different sources. The cost of biomass comes from EA Energy Analyses [20], the cost of gas is based on the cost of gas for non household consumers in Sweden[3] (we assume similar costs in Norway). For the technologies in Table 1, the O&M costs, expressed as a percentage of the investment costs, are respectively: 1, 1.3, 1, 1.3, 2, 0.8, 2.3, 4, 5.5, 1, 1 and 5. For the storage technologies in Table 2, the operating cost is 0. The CO_2 factors of gas and electricity for Norway are based on a report from Adapt Consulting [21] and the CO_2 factor for biomass is based on [22].

The electricity prices for Norway are based on the hourly spot prices for the Oslo region in 2017 from Nordpool.[4] On top of the spot prices, a small retailer fee and the grid charges are added.[5] The prices are rather constant with a fair amount of peaks in the winter and some dips in the summer. This cost structure is close to the actual structure of the electricity price seen by consumers. We assume hourly billing due to its relevance to prosumers and its emergence in Norway.

The irradiance on the horizontal plane and temperatures are obtained and used in the calculations as described in the previous section. The ground station used to retrieve data is Fåvang, situated 50 km to the west of Evenstad. The electric and heat load profiles for the campus are derived from [23]. The load profiles are based on the result of the statistical approach used in these papers and the ground floor area of each type of building on the campus. In addition, the domestic hot water (DHW) and Space Heating (SH) are derived from the heat load based on profiles from a passive building in Finland where both are known [24].

The problems are solved on a Windows 10 laptop with a dual-core CPU (i7-7600U) at 2.8 GHz and 16 GB of RAM. Each case typically has about 450,000 rows, 600,000 columns and 2,400,000 non-zeros. They are solved using the barrier method in Gurobi in about 150 s each.

[3] http://ec.europa.eu/eurostat/statistics-explained/index.php?title=File:Gas_prices_for_non-household_consumers,_second_half_2017_(EUR_per_kWh).png

[4] https://www.nordpoolgroup.com/Market-data1/Dayahead/Area-Prices/ALL1/Hourly/?view=chart

[5] https://www.nve.no/energy-market-and-regulation/network-regulation/network-tariffs/statistics-on-distribution-network-tariffs/

5 Results

The optimization was run several times with different conditions. It was run with a yearly CO_2 balance with and without including the energy system that already exists at Evenstad. When the pre-existing energy system is included, the pre-existing amounts of heat storage, PV, solar thermal and biomass heating (CHP and boilers) represent the minimum possible investments in those technologies for the optimization. The energy systems resulting from those optimizations are presented on Fig. 2.

Both cases are interesting. Indeed the case with the pre-existing technologies included in the optimization allows to know in which technology to invest to move towards being a ZEN for the campus Evenstad while the case that does not include the pre-existing technologies allows to see how it would look like if it was built today from the ground up using the optimization model presented here and the given ZEN restrictions.

A first observation from Fig. 2 is that the technologies already installed (heat storage ST, biomass boiler BB, CHP, battery) do not get additional investments, except for PV which gets a lot of additional investments to meet the ZEN criteria. In addition to the large investment in PV the only additional investment for Evenstad appears to be a heat pump. In the case without any pre-installed technologies the system is quite different. There is still a need for investment in PV, though it is slightly lower and the optimization does not chose to invest in a battery. On the heating part the chosen design uses heat pumps and electric boiler in addition to a heat storage smaller than already installed in Evenstad.

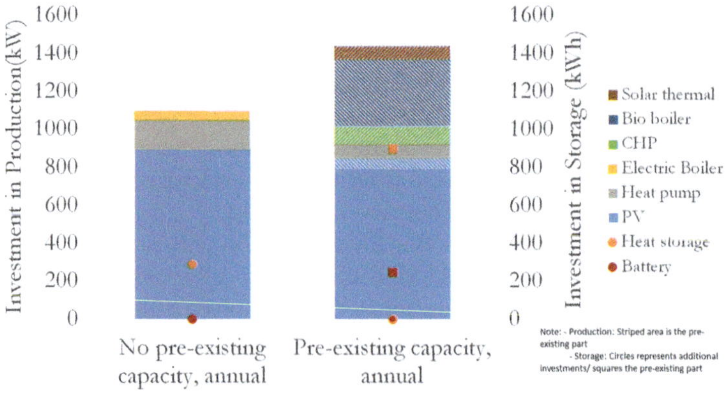

Fig. 2 Resulting energy system

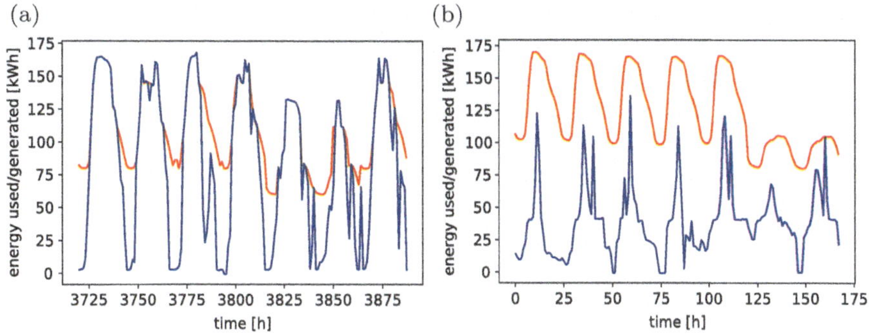

Fig. 3 Self consumed electricity (blue) and total consumption (red) of electricity in the ZEN. (**a**) Summer. (**b**) Winter

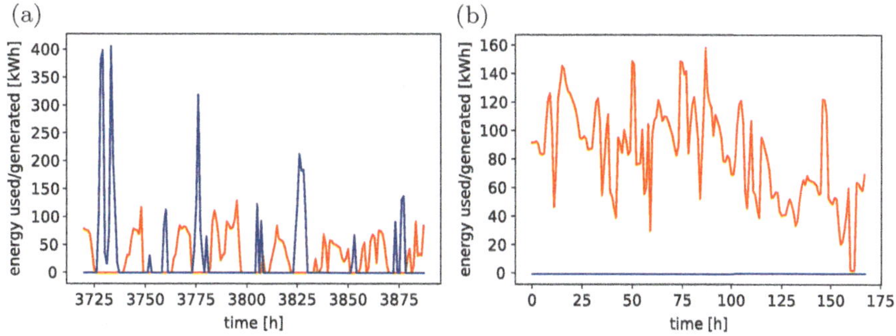

Fig. 4 Import (red) and Export (blue) of Electricity from the ZEN. (**a**) Summer. (**b**) Winter

The results highlight the predominance of PV in the results. This shows that the other possible designs are not cost competitive. Alternative designs, for example relying on biomass CHP, could be incentivized to obtain a better mix of technology. The amount of the incentive could be explored by a sensitivity analysis using this model, but this remains as future work.

On Fig. 3, the self consumption and the total demand of electricity is presented while on Fig. 4 it is the imports (red) and exports (blue) of electricity that are presented. Both figures show a week for the case of the yearly balance and including pre-existing technologies. In the summer the neighborhood produces electricity in excess and needs to send it to the grid. The battery, that is part of the pre-existing technologies, is used but is not large enough to allow for relying on self produced electricity during the night. It is also not large enough to limit the amount of electricity sent to the grid. Figure 4a illustrates this: the exports during the days have highs peaks that represent around four times the night imports in terms of peak power. This has implications on the sizing of the connection to the grid and is especially important in the context of the introduction of new tariffs based on peak power in Norway. Indeed, the introduction of smart-meters enables the use of more complex grid tariff structures. Such tariffs would promote avoiding large peaks in consumption. This may be beneficial to highly flexible neighborhoods such as ZENs and might promote investment in batteries. Investigating the impact of grid tariffs on the design of ZENs remains as future work. Outside of the ZEN context, a positive impact of certain grid tariff designs has been shown on self-consumption and peak electricity import [9]. In the winter, some of the electricity is still self consumed due to the CHP that is part of the pre-existing technologies. This self consumption stays limited and no electricity is exported.

Ultimately, all resulting designs require huge investment in PV to attain the status of ZEN. In those systems, which rely heavily on electricity, heat pumps and electric boilers appear to be the preferred heating solution.

6 Limitations

This study has several limitations, on the methodology and on the case study. For the case study, assumptions were necessary due to the lack of data, in particular for the loads or the insolation (diffuse and direct). For the methodology, the will to limit the use of binary variables meant leaving out constraints such as part load limitations which would be needed to have a better representation of some technologies. In addition, using an hourly resolution leads to an underestimation of the storages and possibly of the heating technologies size. There is a trade off between the solving time and the precision of the results and the resolution needs to be chosen accordingly. Additionally, being deterministic, the model leaves out several uncertainties. Those uncertainties concern the evolution of the price of the technologies, the electricity price or the price of other fuels and the climate conditions. Those can be partially addressed by specifying additional periods in the model. The short-term uncertainties are not included either and induce an overly optimistic operation of the system. Despite those limitations it provides insights in the design methodology that can be used to design the energy system of a ZEN. The choice of CO_2 factors for electricity is also greatly impacting the results and this should be studied in more detail in future work.

7 Conclusion

This paper presented in detail the ZENIT model for investment in Zero Emission Neighborhoods as well as its implementation and the results on a realistic case study of campus Evenstad in Norway, with a focus on methods from the field of operations research. The model is formulated as a MILP, using as few binaries as possible. The Zero Emission constraint complexifies the problematic of designing the energy system of a neighborhood and the long term trends can be accounted for by defining periods. For Evenstad, the results suggest that additional investments, mainly in

PV, are necessary in order to attain the status of ZEN. Investments happen at both levels but mainly at the building level. When the technologies already installed at Evenstad are not included, they are not invested in (except for heat storage). The optimal choice in order to become Zero Emission for Evenstad in the current ZEN framework thus appears to be a massive investment in PV and a heating system fueled by electricity. Further work includes disaggregating the heat part of the model and a more detailed operation part in the optimization.

There are key takeaways for policy makers in this study, in particular for Norway due to the setting of the case study. The results suggest that the Zero Emission constraint used in this study is sufficient to get PV investment without any additional incentive. However, under the CO_2 factor assumptions used in this study, huge investment in PV are made which would be problematic in case of a large-scale application of the concept of ZEN. This suggests the need for incentives in alternative technologies such as biomass CHPs in case the concept of ZEN becomes more common. The methodology presented in this paper can be used to assess such policies and their potential effect on investments in ZEN. Other policies such as the grid tariff structure can also be studied with this model. Finally, the hourly limitation on electricity export from prosumers has recently been replaced in Norway by a tariff on exported electricity. The results of this paper suggest that without this change in policy, ZEN would become even more expensive due to the necessity of large batteries to make the exports more constant. We thus recommend continuing on the path of facilitating the development of the number of prosumers for example with the implementation of capacity grid tariffs. For countries other than Norway, similar methodology can be used to assess the cost and design of ZEN. Further policy recommendations cannot be drawn from this study due to the specificity of the Norwegian electricity mix, that is reflected in its electricity CO_2 factor.

Acknowledgements This article has been written within the Research Center on Zero Emission Neighborhoods in Smart Cities (FME ZEN). The authors gratefully acknowledge the support from the ZEN partners and the Research Council of Norway.

Nomenclature

Indexes (Sets)

$t(\mathcal{T})$	Timestep in hour within year $\in [0, 8759]$
$b(\mathcal{B})$	Building type
yr	Year within period $\in [1, YR]$
p	Period
$i(\mathcal{I})$	Energy technologies, $\mathcal{I} = \mathcal{F} \cup \mathcal{E} \cup \mathcal{HP} \cup \mathcal{S} \cup \mathcal{QST} \cup \mathcal{EST}; \mathcal{I} = \mathcal{Q} \cup \mathcal{G}$:

$f(\mathcal{F})$	Technology consuming fuel (gas, biomass, ...)
$e(\mathcal{E})$	Technology consuming electricity
$hp(\mathcal{HP})$	Heat pumps technologies

$s(\mathcal{S})$	Solar technologies $\in ST, PV$
$qst(\mathcal{QST})$	Heat storage technologies
$est(\mathcal{EST})$	Electricity storage technologies

$q(\mathcal{Q})$	Technologies producing heat
$g(\mathcal{G})$	Technologies producing electricity

Parameters

C_i^{disc}	Discounted investment cost, technology i with re-investments and salvage value [€/kWh]
$\varepsilon_{r,p}$	Discount factor, period p with discount rate r
D	Duration of the study [yr]: $D = P * YR$
P	Number of periods in the study [-]
C_i^{maint}	Annual maintenance cost [% of inv. cost]
$P_{f,p}^{fuel}$	Price of fuel of technology g, period p [€/kWh]
$P_{t,p}^{spot}$	Electricity spot price [€/kWh]
P^{grid}	Electricity grid tariff, period p [€/kWh]
P^{ret}	Retailer tariff on electricity, period p [€/kWh]
η_{est}	Charge/Discharge efficiency of battery est [-]
$\varphi_e^{CO_2}$	CO_2 factor of electricity [g/kWh]
$\varphi_f^{CO_2}$	CO_2 factor of fuel f [g/kWh]
α_{CHP}	Heat to power ratio of the CHP [-]
GC	Size of the neighborhood grid connection [kW]
X_i^{max}	Maximum possible installed capacity of technology i [kW]
$E_{b,t,p}$	Electric specific load of building b in timestep t in period p $[kWh/m^2]$
A_b	Aggregated area of building b in the neighborhood $[m^2]$
$H_{b,t,p}$	Heat specific load of building b in timestep t in period p $[kWh/m^2]$
η_i	Efficiency of technology i [-]
$COP_{hp,b,t,p}$	Coefficient of performance of heat pump hp in building b in timestep t in period p [-]
$\dot{Y}_{max}^{bat}$	Maximum charge/dis- rate of battery [kWh/h]
$\dot{Q}_{max}^{heatstor}$	Maximum charge/discharge rate of heat storage [kWh/h]
$\eta_{t,p}^{PV}$	Efficiency of the solar panel in timestep t in period p [-]
L_i	Lifetime of technology i [yr]
C_{hg}	Cost associated with a heating grid for the neighborhood [€]

Variables

x_i	Capacity of technology i: for $i \in \{f \cup e \cup h \cup s\}$ [kW]; for $i \in \{qst \cup est\}$ [kWh]
$f_{f,t,p}$	Fuel consumed by technology f in hour t [kWh]

$d_{e,t,p}$	Electricity consumed by technology e in timestep t [kWh]
$d_{hp,b,t,p}$	Electricity consumed by the heat pumps hp, in building type b [kWh]
$y^{imp}_{t,p}$, $y^{exp}_{t,p}$	Electricity imported/exported from the grid to the neighborhood at timestep t [kWh]
$y^{exp}_{t,p,g}$	Electricity exported by the production technology g to the grid at timestep t [kWh]
$g^{selfc}_{t,p,g}$	Electricity generated from the technology g self consumed in the neighborhood, timestep t [kWh]
$g^{ch}_{t,p,g}$	Electricity generated from the technology g into the 'prod' batteries at timestep t [kWh]
$y_{\overline{t,p},est}$	Electricity imp/exported by the battery *est* at timestep t [kWh] (gb_exp, gb_imp or pb_exp)
$g_{g,t,p}$	Electricity generated by technology g in timestep t of period p [kWh]
$q_{q,t,p}$	Heat generated by technology q in timestep t of period p [kWh]
$y^{gb\_dch}_{t,p,est}$	Electricity discharged from the 'grid' battery *est* to the neighborhood at timestep t [kWh]
$y^{pb\_ch}_{t,p,est}$	Electricity charged from the neighborhood to the 'prod' battery *est* at timestep t [kWh]
$y^{pb\_selfc}_{t,p,est}$	Electricity to the neighborhood from the 'prod' battery *est*, timestep t [kWh]
$q^{ch}_{t,p}$	Heat "charged" from the neighborhood to the heat storage at timestep t [kWh]
$q^{dch}_{t,p}$	Heat "discharged" from the neighborhood to the heat storage at timestep t [kWh]
$q^{ch}_{t,p}$, $q^{dch}_{t,p}$	Heat "charged"/"discharged" from the neighborhood to the heat storage at timestep t [kWh]
$v^{gb}_{t,p,est}$, $v^{pb}_{t,p,est}$	'grid'/'prod' Battery *est* level of charge at timestep t in period p [kWh]
$v^{heatstor}_{t,p}$	Heat storage level at timestep t in period p [kWh]
$g^{curt}_{t,p}$	Solar energy production curtailed [kWh]
b_{hg}	Binary variable for investment in a heating grid

References

1. M.K. Wiik, S.M. Fufa, J. Krogstie, D. Ahlers, A. Wyckmans, P. Driscoll, H. Brattebø, A. Gustavsen, Zero emission neighbourhoods in smart cities: definition, key performance indicators and assessment criteria. Tech. rep., Research Center on ZEN in Smart Cities (2018)
2. K.B. Lindberg, A. Ånestad, G. Doorman, D. Fischer, M. Korpås, C. Wittwer, I. Sartori, in *Zero Carbon Buildings Today and in the Future* (Birmingham City University, 2014), pp. 145–153

3. K.B. Lindberg, G. Doorman, D. Fischer, M. Korpås, A. Ånestad, I. Sartori, Energ. Buildings **127**, 194 (2016). https://doi.org/10.1016/j.enbuild.2016.05.039
4. P. Gabrielli, M. Gazzani, E. Martelli, M. Mazzotti, Appl. Energy **219**, 408 (2018). https://doi.org/10.1016/j.apenergy.2017.07.142
5. A.D. Hawkes, M.A. Leach, Appl. Energy **86**(7), 1253 (2009). https://doi.org/10.1016/j.apenergy.2008.09.006
6. C. Weber, N. Shah, Energy **36**(2), 1292 (2011). https://doi.org/10.1016/j.energy.2010.11.014
7. E.D. Mehleri, H. Sarimveis, N.C. Markatos, L.G. Papageorgiou, Energy **44**(1), 96 (2012). https://doi.org/10.1016/j.energy.2012.02.009
8. H. Schwarz, V. Bertsch, W. Fichtner, OR Spectrum Quant. Approaches Manag. **40**(1), 265 (2018)
9. H. Schwarz, H. Schermeyer, V. Bertsch, W. Fichtner, Sol. Energy **163**, 150 (2018). https://doi.org/10.1016/j.solener.2018.01.076. http://www.sciencedirect.com/science/article/pii/S0038092X18300975
10. B. Li, R. Roche, A. Miraoui, Appl. Energy **188**, 547 (2017). https://doi.org/10.1016/j.apenergy.2016.12.038. http://www.sciencedirect.com/science/article/pii/S0306261916318013
11. W. Wang, R. Zmeureanu, H. Rivard, Build. Environ. **40**(11), 1512 (2005). https://doi.org/10.1016/j.buildenv.2004.11.017. http://www.sciencedirect.com/science/article/pii/S0360132304003439
12. S. Mashayekh, M. Stadler, G. Cardoso, M. Heleno, Appl. Energy **187**, 154 (2017). https://doi.org/10.1016/j.apenergy.2016.11.020. http://www.sciencedirect.com/science/article/pii/S0306261916316051
13. Y. Yang, S. Zhang, Y. Xiao, Energy **90**, 1901 (2015). https://doi.org/10.1016/j.energy.2015.07.013. http://www.sciencedirect.com/science/article/pii/S036054421500907X
14. H.P. Hellman, M. Koivisto, M. Lehtonen, in *Proceedings of the 2014 15th International Scientific Conference on Electric Power Engineering*, 2014, pp. 269–272. https://doi.org/10.1109/EPE.2014.6839426
15. Å.L. Sørensen, E. Fredriksen, H.T. Walnum, K.S. Skeie, I. Andresen, Zen pilot survey wp4 energy flexible neighbourhoods: Initial plans for thermal and electrical use, generation, distribution and storage. Tech. rep., Research Center on ZEN in Smart Cities (2017)
16. Energinet, D.E. Agency, Technology data for energy plants. Tech. rep., Energinet (2017). https://ens.dk/en/our-services/projections-and-models/technology-data
17. E.T.S.A. Program, Energy supply technologies data. Tech. rep., International Energy Agency (2010–2014). https://iea-etsap.org/index.php/energy-technology-data/energy-supply-technologies-data
18. IRENA, Cost and competitiveness indicators: Rooftop solar pv. Tech. rep., International Renewable Energy Agency (2017). http://www.irena.org/-/media/Files/IRENA/Agency/Publication/2017/Dec/IRENA_Cost_Indicators_PV_2017.pdf
19. IRENA, Electricity storage and renewables: costs and markets to 2030. Tech. rep., International Renewable Energy Agency (2017)
20. C. Bang, A. Vitina, J.S. Gregg, H.H. Lindboe, Analysis of biomass prices: future Danish prices for straw, wood chips and wood pellets. Tech. rep., EA Energy Analyses (2013)
21. A.C. AS, Conversion factors for electricity in energy policy: a review of regulatory application of conversion factors for electricity and an assessment of their impact on eu energy and climate goals. Tech. rep., Adapt Consulting AS (2013)
22. T. Dokka, I. Sartori, M. Thyholt, K. Lien, K. Lindberg, in *Passivhus Norden, The 6th Passive House Conference in the Nordic countries*, 2013
23. K.B. Lindberg, Impact of Zero Energy Buildings on the Power System: a study of load profiles, flexibility and system investments. Ph.D. thesis, NTNU (2017)
24. S.K. Pal, K. Alanne, J. Jokisalo, K. Siren, Appl. Energy **162**, 11 (2016). https://doi.org/10.1016/j.apenergy.2015.10.056

Efficient Operation of Modular Grid-Connected Battery Inverters for RES Integration

Sabrina Ried, Armin U. Schmiegel, and Nina Munzke

Abstract Grid-connected battery storage systems on megawatt-scale play an important role for the integration of renewable energies into electricity markets and grids. In reality, these systems consist of several batteries and inverters, which have a lower energy conversion efficiency in partial load operation. In renewable energy sources (RES) applications, however, battery systems are often operated at low power. The modularity of grid-connected battery storage systems thus allows improving system efficiency during operation. This contribution aims at quantifying the effect of segmenting the system into multiple battery-inverter subsystems on reducing operating losses. The analysis is based on a mixed-integer linear program that determines the system operation by minimizing operating losses. The analysis shows that systems with high modularity can meet a given schedule with lower losses. Increasing modularity from one to 32 subsystems can reduce operating losses by almost 40%. As the number of subsystems increases, the benefit of higher efficiency decreases. The resulting state of charge (SOC) pattern of the batteries is similar for the investigated systems, while the average SOC value is higher in highly modular systems.

Keywords Battery operation planning · Inverter · Energy efficiency · Optimization

S. Ried (✉) · N. Munzke
Karlsruhe Institute of Technology, Karlsruhe, Germany
e-mail: Sabrina.Ried@kit.edu; Nina.Munzke@kit.edu

A. U. Schmiegel
University of Applied Sciences Reutlingen, REFU Elektronik, Reutlingen, Germany
e-mail: Armin.Schmiegel@refu.com

V. Bertsch et al. (eds.), *Advances in Energy System Optimization*,
Trends in Mathematics, https://doi.org/10.1007/978-3-030-32157-4_10

1 Introduction

Grid storage systems on megawatt scale play a vital role for the integration of renewable energies into electricity markets and grids. Several investigations focus on the development of optimized battery operation strategies [1–5]. For several reasons, existing grid storage systems usually consist of multiple batteries and inverters. For reasons of economies of scale, hardware manufacturers offer components in limited number of size classes, allowing lower production costs. Furthermore, the use of modular components supports the systems' scalability. In addition, the size of modular systems can be changed over their lifetime. Moreover, modular systems avoid single points of failure, which leads to higher fault tolerance. Although the modularity of existing grid storage systems is well known, most of the analyses describe the storage system as a single battery combined with a single inverter [1–9]. Few studies are known that analyze the modularity of grid-connected battery systems and the related effect on system efficiency during operation and the influence of these energy losses on the operating strategy of the system [10, 11].

The consideration of this architectural aspect in the model-based analyses of battery operation provides a degree of freedom in optimizing the overall yield of a grid storage system. Figure 1 shows an example for a grid storage system with a high modularity. In this setup, three inverters and batteries are connected to one point of common coupling.

Storage systems operated together with a renewable energy source are most likely not charged and discharged at their nominal power. Figure 2 shows the frequency of the operational inverter power over the course of a year for a large battery that is operated together with a wind farm with the purpose of supporting market integration of wind energy [14].

During 79% of the time, the system is in standby mode. During 5% of the time the system is operated at less than 50% rated power, while during 14% the power rating is between 80% and 100%.

This suggests the importance of energy conversion efficiency not only in full-load operation, but also in standby mode and in partial-load operation. The inverter efficiency, however, is a nonlinear function in terms of power flow (Fig. 3 with

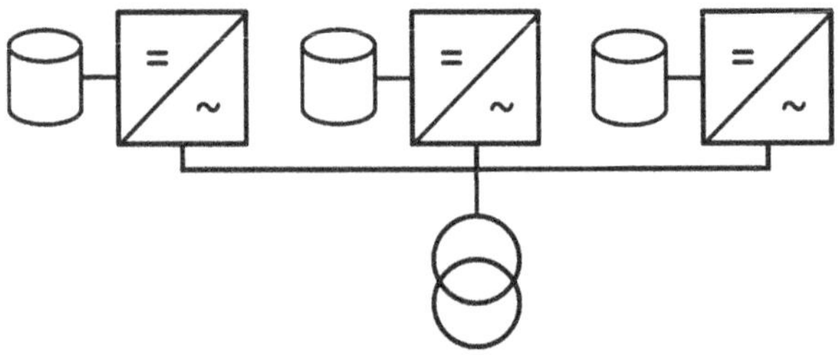

Fig. 1 Example for a grid storage system realized as a combination of three single battery inverter units (N = 3)

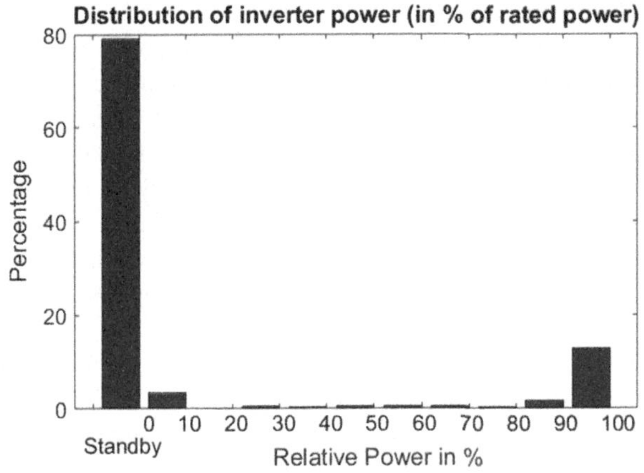

Fig. 2 Frequency distribution of operational inverter power in our case study over one year

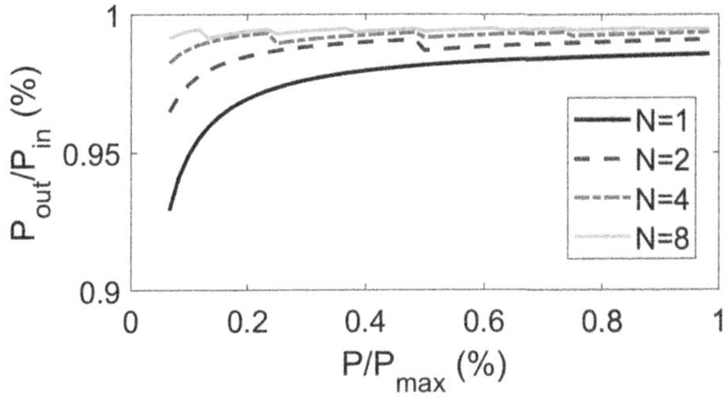

Fig. 3 System efficiency for different number of inverters (N)

N = 1). In partial-load operation, the relative inverter losses are significantly higher in comparison to relative losses at full-load operation. Because of the possibility to use several inverter units, the efficiency is higher in high modularity systems. At low power flows, using one smaller inverter unit can limit the losses. Figure 3 shows the effect of number of inverters on the total efficiency curve. Here, we assume the power rating of the total storage system to be constant. Therefore, the power rating of the inverter and the constant losses scale with $\frac{1}{N}$. The impact on the efficiency is evident up to a number of inverters of 4. Constant losses become less relevant with the increase from 4 to 8 inverters.

Figure 3 provides an overview of the influence on the efficiency of modular battery systems. For operating a system with higher modularity, with discrete battery-inverter pairs, the operating strategy is more complex. Without an optimization strategy, which combines knowledge about power transfer losses, the SOC of the batteries and requests from the grid, suboptimal operating states can be realized, which increases losses. Hence, the operating strategy has a significant influence on the yield of a grid storage system. In this study we avoid choosing an individual optimization strategy by formulating an optimization problem. This optimization problem is formulated in terms of power flows and minimizes the power losses of the system. So, changes in the operating strategy, caused by increasing the number of inverter-battery pairs, are identified by solving the optimization problem.

The target function optimizing the operating strategy of the system is purely technical driven, i.e. we optimize in terms of power losses. Since most of the business models for grid storage systems rely on the power in and outflows of the system for economic evaluation, an optimization in terms of losses is always of benefit for the system operator.

As there is no direct connection between the batteries on the DC-side the energy management needs to take the state of charge of each battery into account. This increases the complexity of the energy management strategy.

This study addresses the questions to which extent the segmentation of the system into multiple battery-inverter subsystems can reduce operating losses and aims at quantifying this effect. Therefore, we develop a mixed-integer linear program that represents the operational strategy of an energy management system of such a modular system as it could be applied in real applications. The mixed-integer linear program determines the system operation for meeting a given schedule for the whole system at the point of common coupling by minimizing inverter losses. As a result, the required energy to be fed into or stored from the grid is allocated over the different battery-inverter-subsystems. The resulting SOC of the batteries is investigated and ideas for further investigations are derived.

Furthermore, the overall system topology has an influence on the power losses and the operating strategy. This study reduces the complexity by looking for identical battery-inverter pairs, which are coupled on the AC-side. A coupling on the DC-side has not been investigated in this study.

This paper is structured as follows. Section 2 describes the mathematical model and presents the data chosen for the analysis. Section 3 presents the results and compares the efficiencies of systems with different degrees of modularity. Section 4 draws conclusions for the design of grid-connected battery systems.

2 Methodology

2.1 Mathematical Model

2.1.1 Power Flow Model

The grid storage system is described as a set of nodes, representing the discrete time steps t with duration of the time steps d, with transfer of energy from one node to another [12, 13]. In this model, each pair of inverter and battery represents a single storage node. The inverter is only described by its influence on the power transfer from the battery into the grid and vice versa. The SOC represents the battery state of charge and interlinks the periods between each other, adding energy flows, both for energy charged and energy discharged during the previous period. N storage systems S_i are connected to one point of common coupling. G^t represents an additional connection to the grid. It serves as a backup when power requests to the storage cannot be met and shall avoid infeasibility of the model.

Hereinafter, a power flow from node A to node B is described as A_B and the efficiency of this transfer is described as $\eta_{AB}(A_B)$. Per definition, transfer losses are assigned to the sink.

The underlying power model is generic and initially independent of the technical implementation of the storage system. Different technologies might add different boundary conditions and parameters to the power flow. All parameters underlying the assumed technical realization are described in 2.1.3 to 2.1.5.

2.1.2 Consideration of Losses

The power transfer losses can be divided into constant, linear, and quadratic terms. This results in the common representation of the inverter efficiency (1):

$$P_{out} = P_{in} - (a_{A_B} + b_{A_B} \cdot P_{in} + c_{A_B} \cdot P_{in}^2) \tag{1}$$

In case of battery storage systems, constant losses a_{A_B} have their origin in auxiliary power, e.g. for the battery management system, active cooling and thermal control of the battery cells, and other subsystems. These losses are independent from the status of the inverter. Some systems can reduce the constant losses during standby operation by switching off subsystems. We consequently assume that the systems considered in our analysis can be turned off, avoiding standby losses during operation, as it is shown to be possible in Munzke et al. [14].

Linear losses b_{A_B} are proportional to the rated power. They are mainly heat losses. Furthermore, we consider battery efficiency. This is reflected by losses that are proportional to the charging and discharging power.

Quadratic losses c_{A_B}represent losses caused by nonlinear saturation effects in the inductance. In this work, the quadratic terms are not taken into account because non-linear losses cannot be clearly observed in all inverter systems [14].

Self-discharge of the battery cells, here represented as power loss of the storage path η_{SS}, is usually very low and thus neglected [14–16]. Moreover, we do not consider the power supply of the battery management system (BMS).

We furthermore do not take into account losses which are independent of the degree of modularity, such as the power consumption of sensors or the climate control. Moreover, we do not consider transformer losses, neglecting the effect of additional power flows, introduced for ensuring model feasibility, on transformer losses.

Inverter loss data is obtained based on a curve fitting approach of the efficiency curve of SMA's 250 kVA inverter "Sunny Central" [17]. The battery loss parameters are approximated based on data measured in Munzke et al. [14].

2.1.3 Parameters and Decision Variables

Table 1 shows the exogenously given parameters used in the model. Table 2 lists the decision variables as well as their lower and upper bounds. Power values are always given in Watt. The penalty parameter p and two decision variables for additional, unplanned power flows to and from the grid (G^t_L and G^t_C) are introduced for reducing deviations from the predetermined schedule and at the same time ensuring the model solvability. The same penalty parameter is applied both for charging from and discharging into the grid.

Table 1 Description of parameters

Parameter	Description	Unit
$G^{max}_{S_i}$	Maximum charge power	W
S^{max}_{iG}	Maximum discharge power	W
κ	Storage power capacity	W
N	Number of inverters	–
T	Number of time steps per optimization	–
d	Duration of one time step t	h
$SOC^{t=0}_i$	Battery state of charge at $t = 0$	Wh
$a_{S_{iG}}$	Constant power losses for discharging (of ith storage)	W
a_{GS_i}	Constant power losses for charging	W
$b_{S_{iG}}$	Linear inverter power losses for discharging	%
b_{GS_i}	Linear inverter power losses for charging	%
$c_{S_{iG}}$	Linear battery losses for discharging	%
c_{GS_i}	Linear battery losses for charging	%
$b_{S_{iS}}$	Linear power losses due to self-discharge of ith storage	%
G^t	Power demand at point of common coupling	W
p	Penalty for additional power flow	–

Table 2 Description of decision variables

Variable	Description	Value Range
$G^t_{S_i}$	Power flow grid to ith storage	$[0, G^{max}_{S_i}]$
S^t_{iG}	Power flow ith storage to grid	$[0, S^{max}_{iG}]$
SOC^t_i	Battery state of charge of the ith storage [Wh]	$[0, \kappa \cdot t]$
γ_{GS_i}	Binary variable for charging of ith storage	$\in \{0; 1\}$
$\gamma_{S_{iG}}$	Binary variable for discharging of ith storage	$\in \{0; 1\}$
G^t_L	Additional power flow to grid	$[0, G^{max}_{S_i} \cdot N]$
G^t_C	Additional power flow from grid	$[0, S^{max}_{iG} \cdot N]$

2.1.4 Target Function

The target function minimizes all losses and penalizes additional power flows (2). The solution is obtained by carrying out one optimization for every hour, i.e. four 15-min time steps at a time.

$$\min Y = \sum_{t=1}^{T}\sum_{i=1}^{N}\Big(a_{S_{iG}}\gamma^t_{S_{iG}} + a_{GS_i}\gamma^t_{GS_i} + (b_{S_{iG}} + c_{S_{iG}})\cdot S^t_{iG} \\ + (b_{GS_i} + c_{GS_i})\cdot G^t_{S_i} + (G^t_L + G^t_C)\cdot p\Big) \tag{2}$$

The target function determines the behaviour of the optimized power management strategy. In a modular system, the given power demand can be met by a subset of inverters. Given the example of two inverters in Fig. 3, only one inverter is used if it can provide the requested power, thus limiting constant losses. When power demand exceeds the single inverter's power rating, the second subsystem is used. In this case, there is no incentive to operate the modules at different power. Therefore, if charge and discharge power requests are sufficiently high, we expect a nearly equal distribution on the SOC.

2.1.5 Constraints

Two binary variables are introduced for charging and discharging ($\gamma_{GS_i}, \gamma_{S_{iG}} \in \{0; 1\}$) in order to ensure that the storage system is not charged and discharged at the same time (3).

$$\gamma_{GS_i} + \gamma_{S_{iG}} \leq 1 \tag{3}$$

The power flow at the point of common coupling is defined by Eq. (4):

$$G^t = G^t_L - G^t_C \\ - \sum_{t=1}^{T}\sum_{i=1}^{N}(G^t_{S_i} + (1 - b_{S_{iG}})S^t_{iG} - a_{S_{iG}}\gamma^t_{S_{iG}}) \tag{4}$$

Equation (5) describes the energy flow for each storage system. It is solved for each hour independently. Therefore, the $SOC^{t=0}$ is set as an external parameter. For the subsequent iterations, the SOC-value of the first time step is set as parameter according to the solution of the optimization of the previous optimization.

$$SOC_i^t = \Big((1 - b_{S_{i_S}}) SOC_i^{t-1} + (1 - b_{G_{S_i}} \cdot c_{G_{S_i}}) G_{S_i}^t \\ - (1 - b_{S_{i_G}} \cdot c_{S_{i_G}}) S_{i_G}^t - a_{S_{i_G}} \gamma_{S_{i_G}}^t - a_{G_{S_i}} \gamma_{G_{S_i}}^t \Big) \cdot d \quad (5)$$

In addition, Eqs. (6) and (7) assure that battery charging or discharging is only realized if the binary variable is equal to 1.

$$G_{S_i}^t - G_{S_i}^{\max} \gamma_{S_{i_G}}^t \leq 0 \quad (6)$$

$$S_{i_G}^t - S_{i_G}^{\max} \gamma_{S_{i_G}}^t \leq 0 \quad (7)$$

Equations (8) and (9) set the boundaries for the power demand at the point of common coupling. These boundaries are calculated outside of and prior to the optimization.

$$G^t \geq -S_{i_G}^{\max} \cdot N \quad (8)$$

$$G^t \leq G_{S_i}^{\max} \cdot N \quad (9)$$

2.1.6 Implementation

The model is implemented in MATLAB, using CPLEX as a solver. The solution is obtained on an hourly basis in 15-min resolution with a MIP gap of 0.5%.

2.2 *Data*

The battery schedule was calculated by Ried et al. [4]. It results from a planned operation of a 50 MW/100 MWh battery which is connected to a 50 MW wind farm. The system participates in the German day-ahead and tertiary control reserve markets. The battery schedule is a time series in 15-min resolution and obtained by a mixed-integer linear program maximizing the contribution margin of the system. For this study, the schedule is scaled to a 2 MW/2 MWh battery system and used as the power demand at the point of common coupling G^t. Since it does not account for detailed losses, the battery used for this study is scaled 30% larger. All assumptions are given in Table 3.

Based on this data, the model is applied to six systems with varying degrees of modularity (Table 4).

Table 3 Parameter values

Parameter	Value
Duration of one time step d	0.25 h
Number of time steps T	4
Storage power capacity κ	2.6 MW
Battery SOC $SOC_i^{t=0}$	1.3 MWh
Maximum charge power $G_{S_i}^{\max}$	2.6 MW
Maximum discharge power $S_{i_G}^{\max}$	2.6 MW
Constant discharging losses $a_{S_{i_G}}$	4.5% $G_{S_i}^{\max}$
Constant charging losses $a_{G_{S_i}}$	4.5% $G_{S_i}^{\max}$
Linear discharging losses (inverter) $b_{S_{i_G}}$	2.1%
Linear charging losses (inverter) $b_{G_{S_i}}$	2.1%
Battery charging losses $c_{G_{S_i}}$	2.5%
Battery discharging losses $c_{S_{G_i}}$	2.5%
Linear losses of self-discharge $b_{S_{i_S}}$	0%
Penalty for additional power flow p	10^{15}

Table 4 Analyzed inverter numbers

No. of inverters	No. of batteries	Inverter size /kW	Battery size /kWh
1	1	2,600	2,600
2	2	1,300	1,300
4	4	650	650
8	8	325	325
16	16	162.5	162.5
32	32	81.25	81.25

3 Results and Discussion

In this section, we compare the yearly losses and resulting battery operation for the different systems, and discuss potential implications for the system layout.

3.1 Losses

Figure 4 shows the losses of systems with varying degrees of modularity relative to the losses of the system with N = 1. In accordance with the assumption shown in Fig. 3, the losses scale with the number of inverter-battery subsystems. By increasing the number of inverters to 32, the operating losses can be reduced by 38%. The gradual decrease of operating losses declines as the modularity increases.

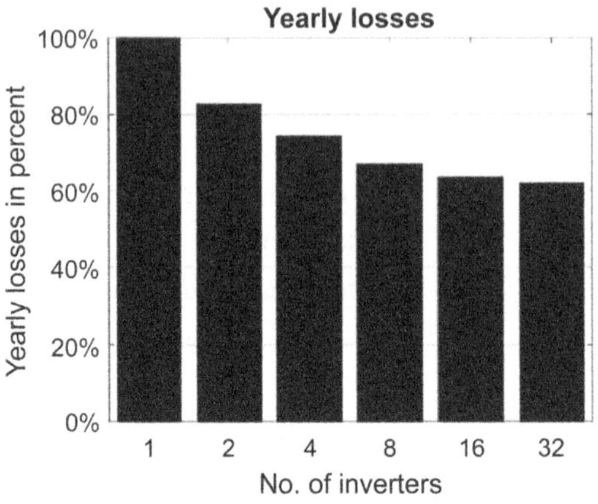

Fig. 4 Yearly relative losses for different numbers of inverters

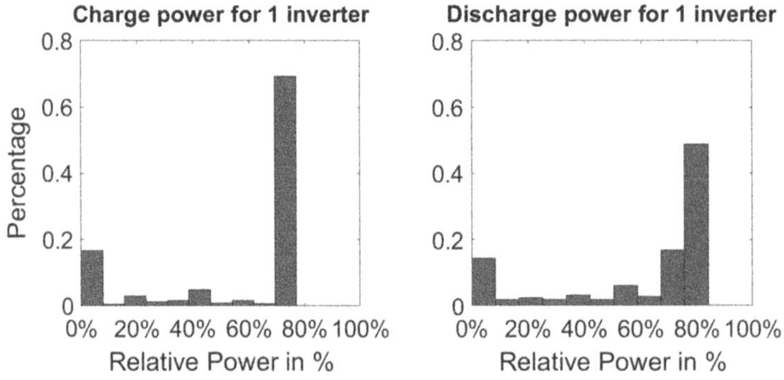

Fig. 5 Charge and discharge power in a single inverter system during operating mode

This effect can be explained if the distribution of charge and discharge power over the course of one year for one and 32 inverters are compared. In a single inverter system, 50–70% of the charge and discharge power, neglecting standby mode, lies between 70% and 80% of the rated inverter power (Fig. 5). While 91% of the energy is charged at a rated power above 70%, 88% of the energy is discharged above 70% power rating. Power below 50% is requested during 29–31% of the time, corresponding to 5–6% of the energy flow. This is the mode of operation in which the individual inverter is less efficient and losses increase.

As the number of inverters increases, the maximum power of each inverter decreases. Figure 6 shows the resulting distribution of charge and discharge power. As expected, the most likely power is the maximum power of the inverter, which corresponds to $\frac{1}{N}$ of the maximum power of the single inverter setup. Only very few events occur, where low power is requested.

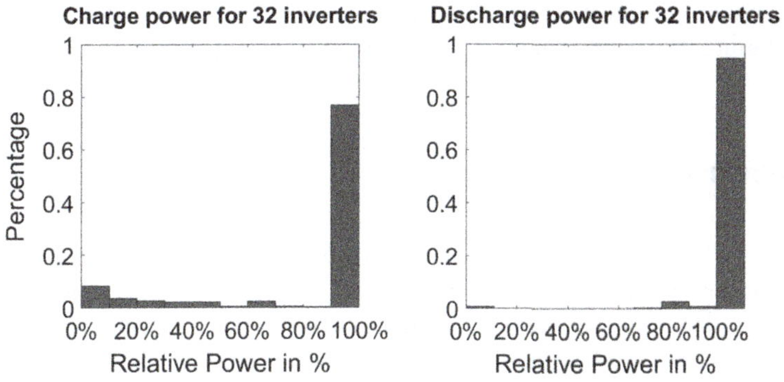

Fig. 6 Charge and discharge power in a system with 32 inverters during operating mode

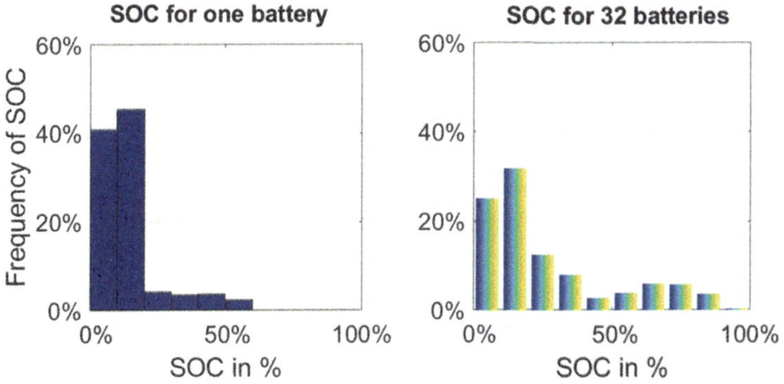

Fig. 7 The distribution state of charge over one year for a realization with one (left) and 32 (right) battery systems

3.2 *Battery Operation*

A difference between the systems with one and 32 inverters is a different SOC-level during the course of the year. Figure 7 shows the distribution of the state of charge over the course of a year for systems consisting of one and 32 sub-systems. In the non-modular system, the battery is operated at an average 13% SOC, while in the high-modularity system the average SOC is 27%. While 86% of the SOC-values in the non-modular system range between 10–20%, 57% of the SOC-values are in this range in the system consisting of 32 inverters and batteries. Due to lower operating losses, less energy is wasted and higher SOC-values occur more often.

Due to a larger number of batteries, there are relatively more downtimes of the batteries in high-modularity systems. While the system consisting of one inverter stands still during 79% of all periods, the sub-systems of the high-modularity systems are in standby-mode during 87% of the time. This suggests a higher redundancy of the system consisting of 32 batteries, which could be beneficial in case of failure, especially when power output must be guaranteed.

4 Conclusions

In summary, we have performed a study on the operating losses of grid-connected battery storage systems on megawatt-scale consisting of several battery-inverter subsystems. Therefore, we applied a mixed-integer linear program determining the optimal operation of all sub-systems for a given schedule for the point of common coupling of a case study where a battery is operated together with a wind farm.

The analysis shows that by increasing the system modularity, the frequency of operating points at low power decreases, resulting in a reduction of operating losses. The higher degree of freedom in modular systems thus allows following a given schedule at lower losses. By increasing the modularity from one to 32 sub-systems, operating losses can be reduced by almost 40%. The effect of modularity on efficiency is most evident for systems consisting of few sub-systems. With increasing modularity, the gradual decrease of operating losses declines. The resulting state-of-charge of the batteries shows a similar operation pattern for the investigated systems. The avoidance of losses in high-modularity systems, however, leads to a shift towards higher SOC levels.

We believe that this work motivates a modular layout of grid-connected battery systems. A higher efficiency should translate into lower operating cost. Moreover, the higher redundancy of modular battery systems can be advantageous particularly in applications that must ensure the system's availability. The related investment cost is another aspect to consider when determining the system design. On the one hand, very large inverter and battery systems might translate into lower investment cost due to less balance of system components. A higher degree of modularity could thus be associated with higher costs. The opposite could also be the case, if the modularity enables the usage of standardized components produced in higher volumes, overcompensating the higher costs for the peripheral components.

Further analyses could address the effect of considering monetary aspects within the target function. Moreover, the penalty parameter ensures model feasibility, but does not fully avoid additional power flows. Both the additional power flows and the choice of the penalty parameter could be further investigated. In order to penalize any deviation from the predetermined schedule, the same penalty parameter was applied both for charging from and for discharging into the grid. However, especially deviations leading to battery discharging might be less critical and sometimes even desirable from energy system perspective, possibly leading to positive revenues. This could be the case if the system participates in additional electricity markets. Consequently, the value of the penalty parameter could be different for positive and negative deviations and should be subject to further research.

Maybe the most interesting potential for further research lies in the operation of the different batteries, the effect on battery ageing and implications on technology choice and system design. Our results show higher average SOC-levels of the batteries in high-modularity systems. The average SOC has an impact on the calendar ageing, so that higher SOC-levels could lead to reduced battery lifetime, at least for most technologies. Another aspect worth investigating is the effect of higher C-rates on battery degradation. Because of allocating the same requested power to fewer batteries with lower capacities, the C-rates might be higher in high-modularity systems. The modular system architecture might thus be more suitable for some battery technologies than for others. The selection of one or more appropriate battery technologies could positively influence battery lifetime.

Lastly, the battery management system determining the operational strategy could be adapted in order to take quadratic losses into account or vary operating patterns, SOC levels, and power requests for different storages. Moreover, asymmetric system topologies would allow for stressing the batteries at different depths of discharge and C-rates, even when running the same schedule. This might be an argument for hybrid storage systems consisting of different battery technologies with different characteristics and furthermore improve cycle ageing of the batteries, another important factor for battery degradation.

References

1. F. Braeuer, J. Rominger, R. McKenna, W. Fichtner, Battery storage systems: an economic model-based analysis of parallel revenue streams and general implications for industry. Appl. Energy **239**, 1424–1440 (2019)
2. M. Gitizadeh, H. Fakharzadegan, Battery capacity determination with respect to optimized energy dispatch schedule in grid-connected photovoltaic (PV) systems. Energy **65**, 665–674 (2014)
3. A. Parisio, E. Rikos, L. Glielmo, A model predictive control approach to microgrid operation optimization. IEEE Trans. Control Syst. Technol. **22**(5), 1813–1827 (2014)
4. S. Ried, M. Reuter-Oppermann, P. Jochem, W. Fichtner, Dispatch of a wind farm with a battery storage, in *Proceedings of the Operations Research Conference*, Aachen (2014), pp. 473–479
5. P. Mokrain, M. Stephan, A stochastic programming framework for the valuation of electricity storage, in *Proceeding of the 26th USAEE/IAEE North American Conference*, Ann Arbor (2006), pp. 24–27
6. L. Hannah, D.B. Dunson, Approximate dynamic programming for storage problems, in *Proceedings of the 28th International Conference on Machine Learning*, Bellevue, Washington (2011)
7. N. Löhndorf and S. Minner, Optimal day-ahead trading and storage of renewable energies—an approximate dynamic programming approach. Energy Syst. **1**(1), 61–77 (2010)
8. Y. Tang, H. He, Z. Ni, J.Wen, Optimal operation for energy storage with wind power generation using adaptive dynamic programming, in *2015 IEEE Power Energy Society General Meeting*, Denver (2015), pp. 1–6
9. A. Das, Z. Ni, T. M. Hansen, X. Zhong, Energy storage system operation: case studies in deterministic and stochastic environments, in *2016 IEEE Power Energy Society Innovative Smart Grid Technologies Conference (ISGT)*, Minneapolis (2016), pp. 1–5

10. S. König, K. Iliopoulos, M.R. Suriyah, T. Leibfried, How direct marketing of a wind power plant affects optimal design and operation of vanadium redox flow batteries, in *Proceeding of the 9th International Renewable Energy Storage (IRES)*, Düsseldorf (2015)
11. M. Schreiber, M. Harrer, A. Whitehead, H. Bucsich, M. Dragschitz, E. Seifert, P. Tymciw, Practical and commercial issues in the design and manufacture of vanadium flow batteries. J. Power Sources **206**, 483–489 (2012)
12. A.U. Schmiegel, A. Kleine, Optimized operation strategies for PV storages systems yield limitations, optimized battery configuration and the benefit of a perfect forecast. Energy Proc. **46**, 104–113 (2014)
13. A.U. Schmiegel, A. Kleine, Upper Economical Performance Limits for PV-Storage Systems, in *Proceeding of the 28th European Photovoltaic Solar Energy Conference and Exhibition*, Paris (2013), pp. 3745–3750
14. N. Munzke, B. Schwarz, F. Büchle, J. Barry, Lithium-Ionen Heimspeichersysteme: performance auf dem Prüfstand, in *Tagungsband des 32. Symposium Photovoltaische Solarenergie, Kloster Banz Bad Staffelstein*, vol. 32 (2017)
15. B.T. Kuhn, G.E. Pitel, P.T. Krein, Electrical properties and equalization of lithium-ion cells in automotive applications, in *Vehicle Power and Propulsion, 2005 IEEE Conference* (2005)
16. A.H. Zimmerman, Self-discharge losses in lithium-ion cells. IEEE Aerosp. Electron. Syst. Mag. **19**(2), 19–24 (2004)
17. Solar Electric Supply, Inc. SMA Sunny Central 250-US Inverter (2018), Available at https://www.solarelectricsupply.com/sunny-boy-sunny-central-250-us-inverters-495. Accessed on 23 Jan 2018

The manufacturer's authorised representative in the EU is Springer Nature Customer Service Centre GmbH, Europaplatz 3, 69115 Heidelberg, Germany. If you have any concerns regarding our products, please contact ProductSafety@springernature.com

Printed and bound by CPI Group (UK) Ltd, Croydon, CR0 4YY
30/09/2026
02225262-0008